TRICERATOPS

A NATURAL HISTORY

TRICERATOPS

A NATURAL HISTORY

Erich Fitzgerald

MUSEUMSVICTORIA
PUBLISHING

Bring extinct creatures back to life and get up close to *Triceratops* on your mobile phone or tablet.

Use the QR code to download your free *Triceratops: Melbourne Museum* augmented reality (AR) app.

1. Scan this QR code with your mobile phone or tablet
2. Tap on the notification to open the link
3. Follow the instructions to download the *Triceratops: Melbourne Museum* AR app
4. You will need to allow the app to access your camera (tap 'OK' when prompted).
5. As you read through this book, keep an eye out for further instructions. Somewhere in this book is a special image that will activate your AR experience!

Plug in your headphones or turn on audio for the full AR experience.

This AR product is compatible with the following devices:

iPhone 8 and above, iPad Pro, iPad 5th generation and above, iPod touch 7th generation using iOS 13 or later

Android devices with a front and back facing camera, using Android 10.0 and later

First published in 2022 by
Museums Victoria Publishing
11 Nicholson St
Carlton, Victoria 3053, Australia
publications@museum.vic.gov.au
www.museumsvictoria.com.au

A catalogue record for this book is available from the National Library of Australia

ISBN: 9781921833571

Design by Simone Hill & Gemma Field
Printed by Adams Print, Australia

1 3 5 7 9 10 8 6 4 2

Museums Victoria acknowledges the Wurundjeri and Boon Wurrung peoples of the Kulin Nations where we work, and First Peoples language groups and communities across Victoria and Australia. Our organisation, in partnership with the First Peoples of Victoria, is working to place First Peoples living cultures and histories at the core of our practice.

CONTENTS

Danny Pearson MP, Minister for Creative Industries

Horridus is the most complete and finely preserved Triceratops *fossils ever found, and one of the most significant dinosaur discoveries in history.*

The Victorian Government is proud to deliver this new world-class attraction and thrilled that Horridus has found a home at Melbourne Museum where they will be on display for all to enjoy.

The acquisition of Horridus the *Triceratops* into our State Collection marks an epoch in the history of our museum and is a coup for Victoria.

While this extraordinary fossil dates back 67 million years, its acquisition and display are focused firmly on the future. Horridus is a major new drawcard for Victoria and will introduce a new generation to our museum, attracting visitors from near and far and supporting groundbreaking scientific research in the process.

Having the world's most complete *Triceratops* here means researchers can begin to uncover the secrets of this magnificent species, further cementing Museums Victoria as Australia's leading museum-based palaeontology research program.

The world-first exhibition, *Triceratops: Fate of the Dinosaurs* highlights our unique Triceratops fossil in an exciting, immersive exhibit that will inspire awe and curiosity in audiences of all ages.

We will remember a time before and after the arrival of Horridus at Melbourne Museum, whose name and remarkable story will inspire visitors for generations to come.

Lynley Crosswell, CEO

It is not at all unusual for museums to collect dinosaur fossils. It is exceptional, however, for a museum to have a specimen of the remarkable quality and significance of Horridus, Museums Victoria's *Triceratops horridus*.

Triceratops is among the most recognisable of all the dinosaurs, second only to *Tyrannosaurus rex* in its ability to inspire excitement among enthusiasts of any age. And yet despite its popularity, there remains a great deal to learn about *Triceratops*, not least the function and purpose of its most distinctive features—the three facial horns and the broad frill that extended beyond its head.

First described from skull fragments collected in the late nineteenth century, *Triceratops* is now one of the most abundant species unearthed by palaeontologists. But there are very few other specimens—and perhaps none—that are as finely preserved as Horridus. And—with 265 separate bones, including an almost entirely intact skull, frill and spinal column—there are currently none that are as complete.

Given its exceptional quality, it was vitally important that, regardless of where it found its forever home, this specimen remained available for scientific study. New computational technologies, including using high-resolution scanning to create detailed 3D digital models, mean that Horridus will be accessible to palaeontological research for generations to come. The exquisitely preserved bones reveal intricate anatomical details, such as the paths of blood vessels and musculature, that are already causing palaeontologists to revisit and rethink what is known about this species.

Beyond palaeontology, Horridus has a great deal to teach us about the fragility of the natural world surrounding us. *Triceratops* came late in the age of dinosaurs, occurring over a 2-million-year period about 67 million years ago. Despite being one of the largest, most awesome herbivores ever to walk the planet, *Triceratops* disappeared in the aftermath of a cataclysmic event that devastated 75 percent of all animal species. The survivors of the age of dinosaurs that live among us are reminders of the prospect of catastrophic biodiversity loss and what that might represent today.

Horridus was acquired for the State Collection with the support of the Government of Victoria, and I express my profound thanks to the State Government of Victoria and the Premier of Victoria, the Hon Daniel Andrews MP, along with Tim Pallas MP, Treasurer of Victoria, Danny Pearson MP, Minister for Creative Industries and Creative Victoria for the generosity of this gift to science, to the people of Victoria and to generations of visitors from around the world who will experience this remarkable fossil.

I would like to express my gratitude to our valued supporters and partners whose contribution has been vital in presenting our *Triceratops* at Melbourne Museum. I am especially thankful to donors Vivian Nadir and Susan Narodowski for their support in memory of their parents, Sam and Nina Narodowski. I am also delighted to welcome Coloursmith by Taubmans as Colour Partner for Museums Victoria and Triceratops: Fate of the Dinosaurs and thank David Nicholls, PPG Taubmans Commercial Director, Australia.

Major Partner VicHealth (Victorian Health Promotion Foundation) is a much-valued supporter of Museums Victoria and I extend my sincere thanks to CEO Dr Sandro Demaio for his continued collaboration. I also wish to thank our Media Partner, Herald Sun, and Penny Fowler, Chairman of the Herald & Weekly Times and Editor Sam Weir, and Tourism Partner V/Line and CEO Matt Carrick.

Lastly, I extend my very sincere thanks to the Museums Board of Victoria and the huge team of Museums Victoria staff who have worked so tirelessly to bring Horridus to their new home at Melbourne Museum.

Philip J. Currie

Professor of Dinosaur Paleobiology, University of Alberta, Canada

Over the years, *Triceratops* has been one of the most influential dinosaurs on my life and career development. It was one of the first plastic dinosaurs that I found in a box of cereal as a prize when I was six years old. It was the dinosaur most often pictured in confrontation with *Tyrannosaurus rex*, most famously portrayed in a spectacular mural by Charles Knight in the Field Museum of Natural History in Chicago. I knew of the mural from a sticker book published by the Audubon Society when I was in Grade 5, but saw the original paintings the same year when we went on a family vacation and passed through Chicago. And once I started my undergraduate studies at the University of Toronto, I discovered the magnificent lithographs of *Triceratops* fossils in the 1907 monograph by Hatcher, Marsh and Lull. I used to sit for hours and leaf through the pages of that magnificent tome!

As one of the first dinosaurs known on the basis of beautifully preserved fossils found in the American West in the late 1800s and put on display in museums in major cities (including New York and Washington), *Triceratops* was one of the handful of dinosaur names known at the dawn of the 20th century. And it was a catchy name that translated into English as 'three-horned face', a simple but elegant moniker that kids everywhere could remember. And as dinosaurs worked their way into books, cartoons, comics, movies, novels and other manifestations of popular culture, *Triceratops* was part of the vanguard. No surprise then that it became one of the most famous dinosaurs worldwide, and still is to this day.

One of the unusual things about the most famous dinosaurs is that many of them (including *Ankylosaurus*, *Triceratops* and *Tyrannosaurus*) were amongst the last-surviving, largest and most specialised dinosaurs of their respective lineages. And yet in many ways they are enigmatic, and there are many unanswered questions about their biology. Although *Triceratops* is known from hundreds of skulls, few skeletons have been collected in association with those skulls. When I worked on the Arctic Islands of Canada I got a sense of why that would happen. Muskoxen, similar to horned (ceratopsian) dinosaurs like *Triceratops*, have very solid skulls; they need to have solid skulls to support the horns that they use for head butting. It is an absolutely amazing thing to hear the loud 'CRACK!' when the heads of two muskoxen collide! As a consequence of their strongly built heads, one often sees muskoxen skulls lying around on the ground in the Arctic with no other evidence of the rest of the skeletons. The bodies get eaten by humans, wolves and other carnivores, but the heads just lie where they fell, and are solid enough that they do not rot away for years. I can imagine that it was much the same with *Triceratops* and other ceratopsian skulls. Another unusual thing about the most famous dinosaurs is that there are so many facts that are contested, or just are still not well enough understood. How many species of *Triceratops* were there? How did it hold its front limbs (was the humerus or upper arm bone held vertically under the body or did it splay sideways)? How long was its tail? Did *Triceratops* lay eggs, and how did it raise its young? Did they move in small groups or massive herds?

Did they use their horns in aggressive behaviour towards each other or to defend themselves against carnivores like *Tyrannosaurus rex*? Given that there are so many specimens of *Triceratops*, that this animal has been studied for more than one and a half centuries (the first partial skeletons were found in 1872), that specimens are beautifully preserved and that it has always been popular, why is there so much we do not know about this iconic dinosaur?

I have been lucky enough to find and excavate skulls and even whole skeletons of ceratopsian dinosaurs. One of my best and most exciting personal discoveries ever was that of the skull and skeleton of a 'baby' *Chasmosaurus*. The animal is very closely related to *Triceratops*, but the specimen I found is even smaller than the smallest *Triceratops* ever recovered as a partial skull and skeleton. The flanks of the body were so well-preserved that you could even see the skin impressions! We published on this beautiful little skeleton in 2016, and again came up with a better understanding of its larger cousin *Triceratops*.

This is a great little book on *Triceratops* because there is something in it for everyone—from the ten-year-old dinosaur enthusiast to the diehard professional like myself. For me it is critical to see the discovery and collection of the specimen documented properly, particularly in such a nice package as this. However, I suspect that most readers will be amazed by how far palaeontologists have been able to tease apart the clues offered by the fossils to interpret the biology of this famous dinosaur. I know that I would have been astounded if I had been able to read this book when I was a ten year old! Yet even though it is an up-to-date account about all of the things we have learned about *Triceratops*, the book emphasises that we still have so much more to learn through further collection and research. There will ALWAYS be room for more people to train and become palaeontologists and have meaningful problems to work on!

My long association with ceratopsian dinosaurs has not dampened my enthusiasm for seeing magnificent specimens. So I jumped when I had the opportunity a few years ago to see the Melbourne *Triceratops* skull and skeleton as preparation began. Unlike other specimens collected and prepared decades ago, the surface textures of the bones had not been obscured by preservatives, paint and dust. And there were many parts of other *Triceratops* skeletons that I had never been sure of how much restoration had been included (like the hands, feet and tail). Like most museum exhibits, there is never enough room to include all the information on the label. And that is where this book comes in handy as the perfect companion for the display. For me the Melbourne *Triceratops* is my greatest experience with this animal since I found that first plastic representation in the box of cereal when I was six!

INTRODUCING

TRICERATOPS

Imagine an animal unlike anything alive on Earth today.

As large, or larger, than an African elephant, this animal was 7 metres long, stood 2.5 metres tall and weighed well over 6000 kilograms. It had a body covered in large scales, a slender tail, wide hips, a great barrel-shaped mid-section and a short neck. Its huge head had a sharp beak at the front and a broad, bony frill fanning out from the back. In-between the beak and frill were three horns: a small triangular horn on its nose and one long horn on each brow. It is the Greek words for 'three-horned face' that combine to make this iconic animal's name—*Triceratops!*

Triceratops was one of the most abundant herbivorous dinosaurs in western North America at the end of the Cretaceous Period, 68 to 66 million years ago. It certainly would have witnessed the devastating environmental changes that caused its extinction and the end of its world.

Today, *Triceratops* is one of the most legendary dinosaurs in popular culture. In our minds, *Triceratops* hatches from eggs, chomps through plants and duels with its horns, defending against its nemesis *Tyrannosaurus rex*. Is this what *Triceratops* and its world were really like? The latest scientific research, including amazing fossil discoveries, reveal that *Triceratops* was more wondrous and intriguing than we ever imagined.

Triceratops, Charles R. Knight, 1901, Smithsonian Institution

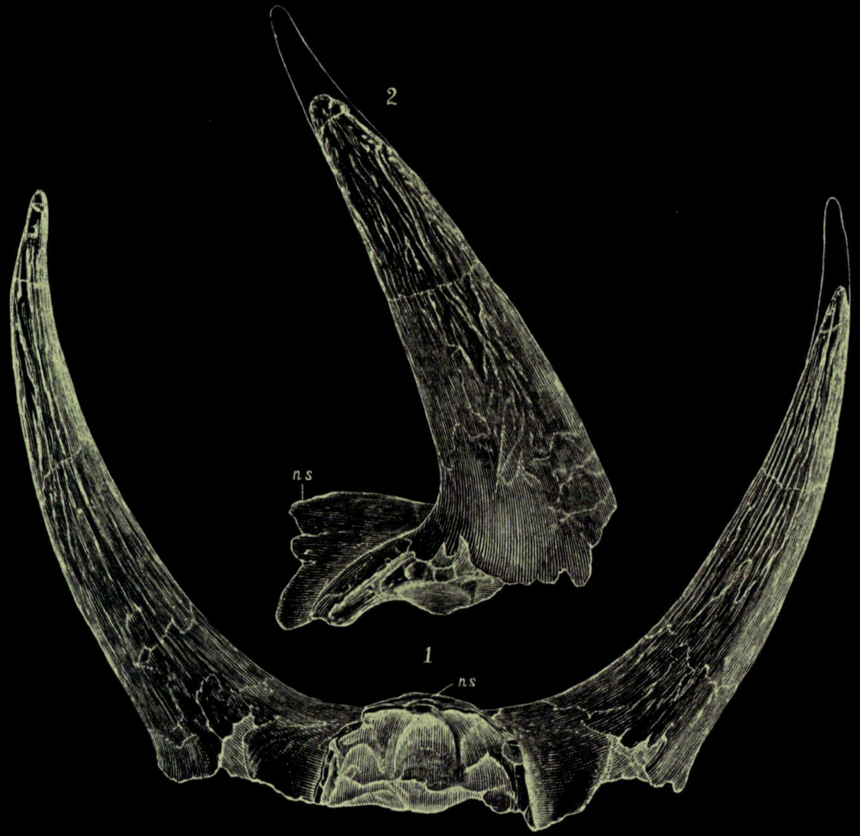

Discovered in 1887, this pair of 60-centimetre-long fossil horn cores were named *Bison alticornis*.

Discovery and Mistaken Identity

Triceratops is known today as the most famous horned dinosaur, but when the first undoubted *Triceratops* fossil was found near Denver, Colorado in 1887, palaeontologists thought it was a giant extinct bison! This makes sense, because that very first fossil was just a pair of enormous horn cores that closely resembled those of the living American bison. Because of this resemblance, palaeontologist Othniel Charles Marsh—who made the discovery—named these horn cores *Bison alticornis*. History was made two years later when a more complete fossil skull was found by ranchers in Wyoming.

This remarkable fossil was clearly a strange new dinosaur. It had a horn core on its nose, but also a pair of long horn cores on its brow. This revealed the true identity of the earlier 'bison' fossil. In August of 1889, Marsh introduced the name *Triceratops* to the world.

Over the next few years, Marsh sent legendary field palaeontologist John Bell Hatcher out on multiple forays into the remote American West, where he found dozens more fossils. In 1891, these additional bones enabled Marsh to present his idea of what the whole skeleton of *Triceratops* looked like. Since then, as more fossils of *Triceratops* have been found and studied, earlier ideas on its anatomy have been tested. This has led to changes in how palaeontologists think this dinosaur looked, moved and lived in its ancient environment.

Changing Times, Changing *Triceratops*

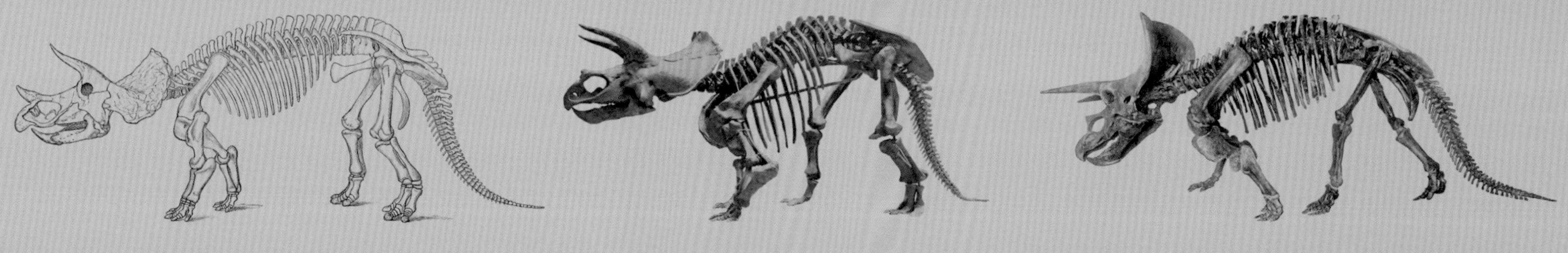

1891

The very first attempt to reconstruct the skeleton of *Triceratops* included too many vertebrae between the shoulders and hips. The forelimbs were illustrated straight under the body and the hindfeet were given three toes when they should have had four. These inaccuracies are understandable because no complete skeleton of *Triceratops* had yet been found.

1905

The Smithsonian Institution assembled the first *Triceratops* skeleton on public display to include real fossils, and the correct (shorter) backbone length between the skull and hips. For the arm bones to articulate at the shoulder joint, the forelimbs could not be straight like the 1891 version. Instead, the elbows stuck out to the sides, making the forelimbs sprawled.

1923

Eighteen years later, the American Museum of Natural History built a *Triceratops* skeleton with the forelimbs even more sprawled out on either side of the body, like those of a lizard. This forced the backbone to slope steeply down to the low-slung skull. Today, palaeontologists think the forelimb posture looked closer to the 1905 attempt.

Three-Horned Face

Triceratops was one of the last horned dinosaurs of the family Ceratopsidae, evolving at the very end of the Cretaceous Period—about 68 million years ago—in a thin slice of geologic time called the Maastrichtian Stage. Fossils of *Triceratops* have only been found in sedimentary rocks of this age in a handful of places across western North America: in the provinces of Alberta and Saskatchewan in Canada; and in Colorado, Wyoming, South and North Dakota, Montana and possibly Utah in the United States of America. One body of sedimentary rock layers in Montana, known as the Hell Creek Formation, is especially rich in dinosaur fossils.

For nearly two million years, *Triceratops* was the most common dinosaur throughout its range. We know this because wherever you find dinosaur fossils of this age in these parts of North America, you find *Triceratops*—and lots of them. You can barely walk across the Hell Creek Formation without tripping over a *Triceratops* bone.

For much of the Cretaceous Period, the Western Interior Seaway ran up the middle of North America, splitting it into a western and an eastern landmass. By the time of *Triceratops,* this seaway had become even shallower, reduced to large inland seas and lakes.

One palaeontologist claimed to have seen fragments of 500 *Triceratops* skulls during seven years of fieldwork!

Recent research has calculated that Triceratops *makes up about half of all the dinosaur remains in the Hell Creek Formation.*

Although bone fragments, single bones, skull parts and even nearly intact skulls of *Triceratops* might be surprisingly common, full skeletons of *Triceratops* are extremely rare. In fact, more skeletons of *Tyrannosaurus rex* have been found! In the 134 years since its first discovery, fewer than 10 examples of *Triceratops* that preserve more than half of the bones in the skeleton have been collected, and none of them are close to complete. This means that there are still many unanswered questions about one of the world's favourite dinosaurs.

With unique insights from one special fossil, this book will help you get to know *Triceratops* in a whole new way. Discovered on a Montana ranch in 2014 and now permanently housed in Melbourne Museum, this specimen is a magnificent, nearly complete skeleton of one individual *Triceratops*. We will explore what we know about the life and times of *Triceratops*, the discovery of this scientifically priceless fossil and how its story reveals the fate of the dinosaurs. To begin with, we must go back to the evolutionary roots of *Triceratops* and the rise of the horned dinosaurs.

The layers of sandstone and mudstone in the Hell Creek Formation erode to form these badlands in Montana.

EVOLUTION OF

TRICERATOPS

Triceratops is one of many species of herbivorous horned dinosaurs called ceratopsids, classified in the family Ceratopsidae. All ceratopsids had large bodies, brow horns, frills on their massive heads, teeth with two roots and they walked on four legs.

Ceratopsians belong in the Ornithischia, a major group of beaked and herbivorous dinosaurs that also includes *Stegosaurus*, *Ankylosaurus*, hadrosaurs and the closest relatives of ceratopsians—the Pachycephalosauria or 'bonehead' dinosaurs.

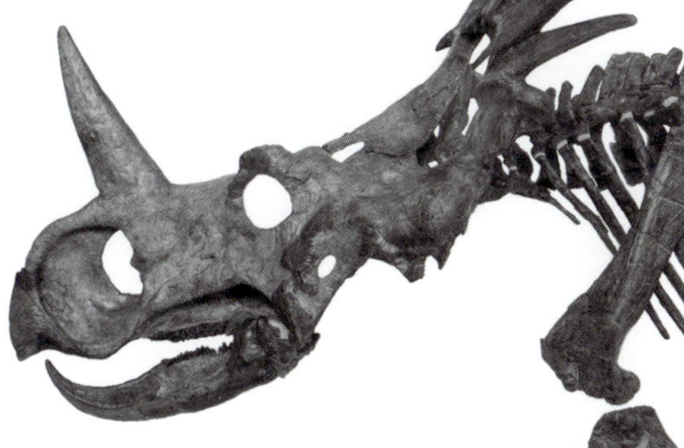

Although pachycephalosaurs (*Pachycephalosaurus*, at left) have very different skulls to horned dinosaurs like *Styracosaurus* (at right) they are the closest relatives of the Ceratopsia.

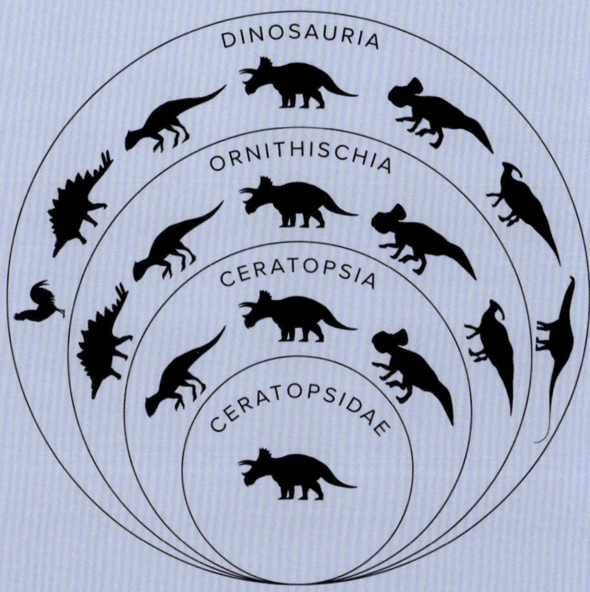

Ceratopsidae

The Ceratopsidae are one part of a larger dinosaur group called the ceratopsians or Ceratopsia. So, Ceratopsidae are *also* ceratopsians, but not all ceratopsians are ceratopsids. This may sound confusing, so this diagram above helps explain it. A simple way to remember the difference between the definition of ceratopsid and ceratopsian is that a ceratopsid is a ceratopsian with a frill *and* brow horns *and* teeth with two roots.

The First Ceratopsians

The most ancient fossils of ceratopsians have been found in China in rocks dating from the Jurassic Period, between 167 and 160 million years ago. One of these Jurassic ceratopsians, *Yinlong downsi,* is known from nearly complete skeletons that provide a good idea of what the earliest relatives of *Triceratops* were like. *Yinlong* was a small dinosaur, between 1 and 2 metres long. Its hindlimbs were twice as long as its forelimbs, showing that the earliest ceratopsians were bipedal. Unlike later ceratopsians, *Yinlong* did not have a frill, but it did have a bone at the tip of the upper jaw that supported the beak. This bone is called the rostral, and is unique to ceratopsians.

The Jurassic ceratopsian *Yinlong downsi* hides from the early tyrannosaur *Guanlong wucaii*.

Ceratopsians of the Cretaceous

The Cretaceous Period was when ceratopsian evolution took off, diversifying into a stunning array of species and expanding out from Asia to other parts of the world, arriving in North America by about 106 million years ago. At that time, all known ceratopsians were still bipedal and small, between the size of a cat and a sheep.

Psittacosaurus from Asia had a short and deep skull with a parrot-like beak and wide cheekbones, but no horns or frill. *Auroraceratops* from China is one of the earliest ceratopsians to have a short frill and small horns called epijugals at the tips of its wide cheekbones—key features of all later-evolving ceratopsians, including *Triceratops*.

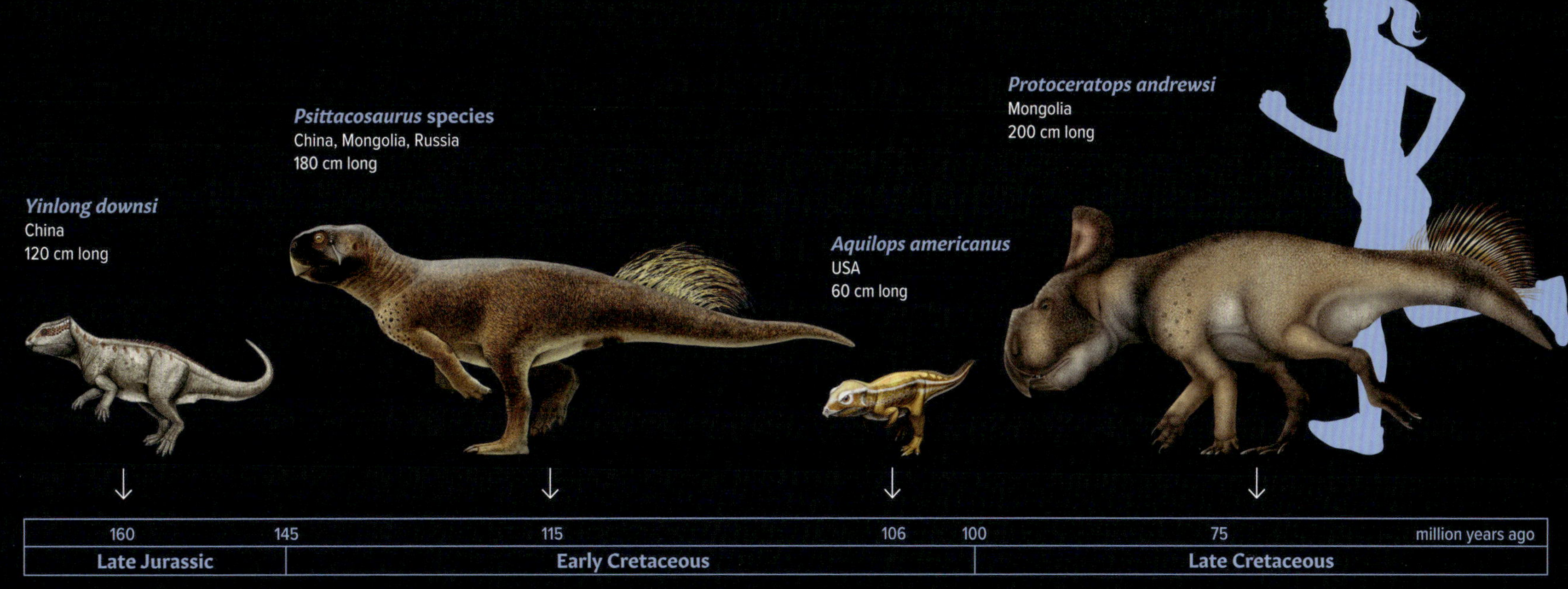

The skull and artist's impression of the head of *Protoceratops andrewsi*.

The Late Cretaceous, between 100 and 66 million years ago, was the heyday of the ceratopsians. This is the chunk of time from which most ceratopsian fossils have been found. The Protoceratopsidae, including *Protoceratops* from the Gobi Desert of Mongolia, was one of the earliest evolving lineages to show the shape—and size—of ceratopsians to come. *Protoceratops* adults were quadrupedal, up to 2 metres long, and had nearly one-metre-long skulls with a well-developed frill. *Protoceratops* would have looked a bit like a miniature *Triceratops*, but with one missing ingredient: horns.

Enter the Horned Dinosaurs—Ceratopsidae

The Ceratopsidae—the horned dinosaurs—are almost entirely known from fossils found in western North America. It was there, over the last 14 million years of the Cretaceous, that they evolved into more than 40 herbivorous species that were the size of rhinoceros and elephants, or in some cases even larger. They had jaws bearing dozens of advanced shearing teeth and specialised forelimbs for a heavy quadrupedal lifestyle. Each species had a different frill size and shape, and the number, size and arrangement of horns, knobs, lumps and spines on the skull varied too. Palaeontologists use these unique combinations of skull features to identify ceratopsid fossils and what species they belong to.

The wonderfully bizarre horns and frills of ceratopsids look like fearsome medieval lances and shields, but it's likely that they were mainly used for visual displays or signals to other ceratopsids, with a backup role as defensive or attacking weapons when needed.

Resting *Zuniceratops* watch on as the little tyrannosaur *Suskityrannus* passes by.

A ceratopsian close to the beginnings of ceratopsids is *Zuniceratops christopheri*, found in 92-million-year-old rocks in the southwest of the United States of America. Although much smaller than ceratopsids at only the size of a sheep, *Zuniceratops* had a large bony frill and long brow horns. *Zuniceratops* shows us that long brow horns were probably a feature of the first ceratopsids and that ceratopsids may have first evolved in North America. Between *Zuniceratops* and the oldest known ceratopsid fossils there is a gap of about 10 million years. What happened in horned dinosaur evolution during those 10 million years is a question that can only be answered with more fossil discoveries.

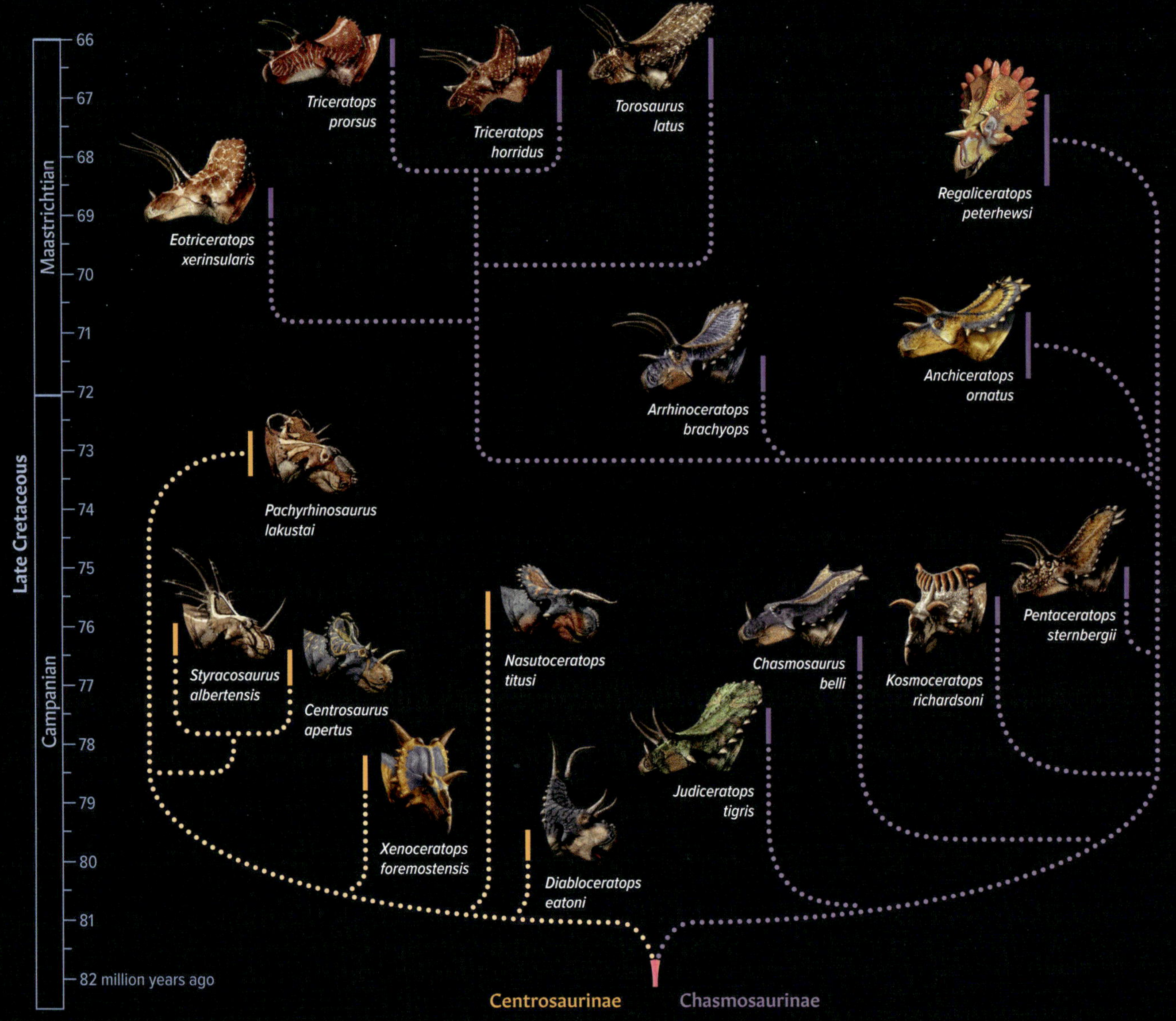

Centrosaurines and Chasmosaurines

By the time the first ceratopsids are found in the fossil record they had already split into two major branches: the Centrosaurinae or centrosaurines, and the Chasmosaurinae or chasmosaurines.

The 80-million-year-old *Diabloceratops* ('devil-horned face') is one of the earliest known centrosaurines and had the short, high snout and short, highly decorated frill that set centrosaurines apart from chasmosaurines.

Later-evolving centrosaurines such as *Centrosaurus* ('pointed lizard') had long, pointed horn cores on their noses and short horn cores on their brows, with *Styracosaurus* ('spiked lizard') also having as many as six or seven long spikes on the rear edge of its frill for good measure.

The chasmosaurines evolved enormous skulls with longer snouts and more complex nostril openings. They have long horn cores on their brows and short horn cores on their noses, plus frills that are long and simply decorated. These massive skulls—the largest of any land animals in history—were carried by gigantic bodies, with many chasmosaurines exceeding the size of the largest known centrosaurines.

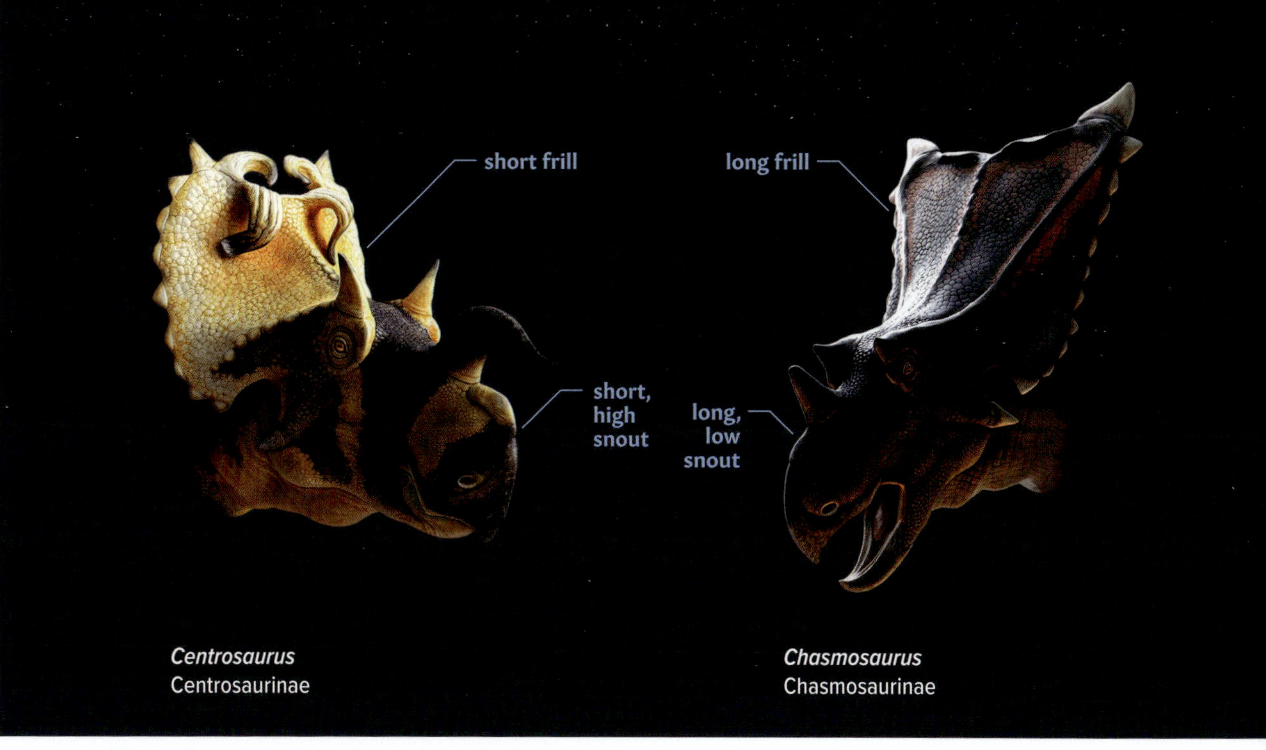

Centrosaurus
Centrosaurinae

Chasmosaurus
Chasmosaurinae

Centrosaurinae

- *Pachyrhinosaurus canadensis*
- *Centrosaurus apertus*
- *Styracosaurus albertensis*
- *Nasutoceratops titusi*

Chasmosaurinae

- *Eotriceratops xerinsularis*
- *Triceratops horridus*
- *Pentaceratops sternbergii*
- *Chasmosaurus belli*

During the Campanian Stage, between 80 and 71 million years ago, both centrosaurines and chasmosaurines evolved into many different species. By 68 million years ago, in the Maastrichtian Stage, the only ceratopsids left seem to have been chasmosaurines. Why did the centrosaurines go extinct while the chasmosaurines were able to hang on?

Chasmosaurines (*Chasmosaurus*) and centrosaurines (*Styracosaurus*) coexisted in a rich community of large herbivorous dinosaurs in the Campanian of Alberta, Canada, 76 million years ago.

Palaeontologists have only found a few fossils that give a glimpse of what was happening in ceratopsid evolution at that time.

One of those fossils is the unusual chasmosaurine *Regaliceratops* ('royal horned face'), which was found in rocks that were between 68.5 and 67 million years old in Alberta, Canada. *Regaliceratops* has a large horn core on its nose, small horn cores on its brow and a 'crown' of large triangular bones around the edge of its frill, which make it *look* like a centrosaurine.

So, just before the end of the Cretaceous Period, some chasmosaurines evolved in surprisingly similar ways to the extinct centrosaurines, perhaps filling the ecological space they left vacant.

Meanwhile, other chasmosaurines evolved larger bodies, longer brow horns and thicker frills. Among them were the ancestors of *Triceratops*.

The unusually ornamented chasmosaurine *Regaliceratops*.

Origin of *Triceratops*

For such an iconic dinosaur, *Triceratops* was a latecomer. Its oldest fossils are between 69 and 67 million years old—that's about 184 million years after the beginning of the 'Dinosaur Era' and just two or three million years before its end! We don't know what the direct ancestor of *Triceratops* was, but an early relative called *Eotriceratops* ('dawn three-horned face') lived 69 million years ago in what is now Alberta, Canada. *Eotriceratops* shares many skull features with *Triceratops* and may have grown to the same size or even larger, more than 8 metres long.

The ceratopsids that could be most closely related to *Triceratops* are *Nedoceratops* and *Torosaurus*. *Nedoceratops* ('insufficiently horned face') is known from one skull found in the same rocks as *Triceratops* and *Torosaurus* fossils in Wyoming. The *Nedoceratops* skull looks a lot like that of *Triceratops*, but the horn core on its nose is low and rounded, the horn cores on its brow point nearly vertically and there are holes in its frill that *Triceratops* skulls lack. Some scientists think *Nedoceratops* is just a strange, maybe diseased, example of *Triceratops*, while other research has shown that *Nedoceratops* is a closely related, but different, genus from *Triceratops*.

Torosaurus ('perforated lizard') gets its scientific name from the pair of large holes in its extremely long frill bones. *Triceratops* frills lack these holes. *Torosaurus* lived at the same time and in the same place as *Triceratops*, but its fossils are far less common. A few scientists have argued that *Torosaurus* skulls are elderly *Triceratops*, which would mean that there is no such thing as *Torosaurus*—they're all senior *Triceratops!* Other scientists think it is unlikely that immature *Triceratops* skulls could grow into the very different skull anatomy seen in *Torosaurus* fossils. As more fossils are discovered and more research is done, it seems increasingly likely that *Torosaurus* is a genuinely different dinosaur to *Triceratops*.

The immense skull of *Torosaurus*, showing one of the pair of holes in the frill that inspired its name.

Triceratops prorsus skulls have a longer, forward pointing nose horn core, and relatively short brow horn cores.

Triceratops horridus skulls have a short, upward pointing nose horn core, and relatively long brow horn cores.

Triceratops Species

Sixteen different species of *Triceratops* have been named, but over the years these various species have been sorted, combined or rejected so that now there are only two accepted species of *Triceratops: Triceratops horridus* and *Triceratops prorsus*. Palaeontologists can usually tell these two species apart by differences in the size and shape of the upper beak and nose horn core as well as the length of the horn cores on their brow. T*riceratops horridus* and *Triceratops prorsus* had bodies that were about the same size. The Melbourne Museum *Triceratops* has been identified as belonging to the species *Triceratops horridus*.

Imagine that you could travel back to the Late Cretaceous to do some dinosaur-watching in the environment of the Hell Creek Formation. Which species of *Triceratops* would you see? This would depend on when you were visting what would one day become the Hell Creek Formation. Scientific research has shown that *Triceratops prorsus* existed for only the last 300,000 years or so of the Cretaceous Period. In fact, *Triceratops prorsus* seems to have been the one and only species of *Triceratops* alive during the final years of the Cretaceous. The last *Triceratops* alive on Earth were *Triceratops prorsus*.

Does this then mean *Triceratops horridus* was the first and earliest species of *Triceratops?*

Perhaps not. By comparing lots of different skulls of *Triceratops* found in the bottom (oldest), middle and top (most recent) strata of the Hell Creek Formation, palaeontologists have discovered that fossils from the oldest rocks look like *Triceratops horridus*, but they are not quite the same. In fact, it is possible that these oldest *Triceratops* fossils are a new, unnamed species. Some scientists think that *Triceratops horridus* could have evolved from this earlier *Triceratops*, then *Triceratops horridus* evolved into another later species of *Triceratops*, which then finally evolved into the last species of all: *Triceratops prorsus*. Research is continuing to unravel the evolution and classification of ceratopsid dinosaurs, so we should expect more twists in the tale of *Triceratops*.

For now, we know that Triceratops horridus *was an earlier species of* Triceratops, *not the last.*

Its fossils have only been found in rocks between 68 and 66.5 million years old in western North America. It is the environment and ecosystem of that time and place—the world of *Triceratops horridus*—that we shall explore next.

Triceratops in its lowland environment.

THE WORLD OF

TRICERATOPS

If we could travel back to 67 million years ago and look down on North America from orbit, its position on the globe would look much like it does in the present day.

Strange to our eyes would be the connected inland seas that extended from the northwest down through the interior of the continent. This chain of linked bodies of water was the last remnant of the vast and shallow Western Interior Seaway that once divided North America into western and eastern landmasses. The sand, mud and clay deposited at the bottom of rivers, ponds and swamps along the western coastlines of these inland seas would become the rocks establishing the lower half of the Hell Creek Formation.

Western North America in the latest Cretaceous, 67 million years ago.

Environment of the Hell Creek Formation

The Hell Creek ecosystem was a humid coastal lowland with rivers meandering through a patchwork of swamps, woodland and old-growth forest dotted with ponds. This was a greenhouse world, with the Montana region having a subtropical climate similar to present-day Florida or northern Australia. The year would have been divided into a wet season, with annual rainfall between 900 and 1200 millimetres causing floods, and a dry season, with regular forest fires indicated by abundant fossil charcoal. The plants and animals that shared this world with *Triceratops horridus* were a mix of the familiar and bizarre, forming a biodiverse ecosystem that showed no signs of decline.

LEFT TO RIGHT

Triceratops

Damselfly

Salamander

Thescelosaurus (ornithopod dinosaur)

Baenid turtle

Anzu (beaked theropod)

Didelphodon (metatherian mammal)

Multituberculate (small mammal)

Plants

Hundreds of fossil plant species have been found throughout the Hell Creek Formation, the most common being the leaves of angiosperms (flowering plants). Ferns and mosses were present as low-growing ground cover, but there is no evidence of tree ferns. Grasses were nowhere to be seen. Ginkgoes and cycads were present, but their fossils are rarely found. In and around the swamps were stands of swamp cypresses (*Taxodium*) and horsetails grew along the edges of waterways. The subtropical woodlands were open habitats, home to the relatives of the palmetto palm (*Sabalites*), a close relative of ginger (*Zingiberopsis*) and the 20-metre-tall *Dryophyllum*, a tree from the walnut family. Other canopy trees that grew in the moist soil of forests between rivers were dawn redwoods (*Metasequoia*), plane trees (*Erlingdorfia*) and katsuras (*Cercidiphyllum*).

Fish

The flooding of North America by the Western Interior Seaway made it easier for ocean fish, including a variety of sharks and rays, to enter and adapt to fresh water in the Hell Creek ecosystem. Sharks included *Lonchidion*—a small species of hybodont shark that survived from the Triassic—nurse sharks and abundant freshwater wobbegongs. The small, flat crushing teeth of the extinct freshwater guitarfish *Myledaphus* account for more than half of all shark and ray fossils found in the Hell Creek Formation. River channels were home to more than 20 species of bony fish, many of which were species of sturgeon, paddlefish and gars whose descendants still inhabit North America's rivers. Bowfins were common, with the extinct *Melvius thomasi* growing to a monstrous 1.6 metres long. At the other end of the size spectrum were deep-bodied 'bony-tongued' fish fewer than 20 centimetres long. These tiddlers could have been a meal for the extinct pike *Estesesox*, a torpedo-shaped ambush predator.

Amphibians

Riverbeds, backwaters and bogs were crawling with amphibians, including at least nine species of salamanders in the lower Hell Creek Formation. Their most frequently found fossil remains are jaws and vertebrae, so it is fortunate that these bones can be used to identify the different types of salamanders. The most common species lived in water, had very short legs and long bodies. *Scapherpeton* looked like the living hellbender, grew to 100 centimetres long and preyed on a wide range of small animals. The giant extinct siren *Habrosaurus* had only two tiny arms, a 160-centimetre-long eel-like body, and hunted insects, water snails and small fish in rivers. The 20-centimetre-long *Albanerpeton* belonged to a mysterious extinct group of scaled amphibians. It inhabited the moist soil around swamps and was a sit-and-wait predator that fired a sticky ballistic tongue at insects that strayed too close.

LEFT TO RIGHT

Myledaphus (guitarfish)

Pachycephalosaurus (dome-headed dinosaur)

Galagadon (wobbegong shark)

Champsosaurus (aquatic reptile)

Gar

Edmontosaurus (duck-billed dinosaur)

Mammals

Well before the 'Age of Mammals' had even started, mammals evolved into a surprising variety of body forms and lifestyles. Sixty-seven million years ago, the Hell Creek ecosystem was inhabited by at least 30 different species of mammals belonging to three major groups: the now-extinct multituberculates, early relatives of living marsupials called metatherians and the eutherians—the branch of the mammal evolutionary tree that humans and all other placental mammals are part of.

Despite their present-day dominance, the eutherians were the least diverse group of mammals in the Hell Creek Formation, being mostly mouse- or rat-sized carnivores dwelling in the undergrowth or canopy.

The multituberculates were a species of ground-dwelling diggers and burrowers that ranged from the size of a mouse (e.g. *Mesodma*) to a cat (e.g. *Meniscoessus*). They had strong jaws, chisel-like incisor teeth and grinding molars that they used to feed on insects, plant roots and leaves. Many of the pouched metatherians were brown rat-sized tree climbers, feeding on fruits and insects with small, sharply pointed teeth.

A big exception was *Didelphodon vorax*, which at 5 kilograms in weight and 1 metre in length was one of the heftiest Mesozoic mammals. *Didelphodon* had long canine teeth, strong molars and heavily-muscled jaws, showing it was a powerful predator and scavenger that could have eaten any animal as big as itself—including small dinosaurs!

Didelphodon vorax

Reptiles

After fish, turtle remains are the most commonly found fossils in the Hell Creek Formation. At least 26 species have been identified so far, making the Hell Creek ecosystem a turtle 'hot spot'. The dominant group of aquatic turtles were the extinct long-tailed baenids, which mostly inhabited ponds, although the 35-centimetre-long *Saxochelys gilberti* preferred rivers. The carnivorous trionychid or softshell turtles, which survive in the present day, were abundant in the rivers, swamps and ponds. Some species, such as *Gilmoremys lancensis,* were small, just 35 centimetres long, but *Axestemys infernalis* had a shell 75 centimetres in length. The true giant among Hell Creek turtles was the extinct *Basilemys sinuosa*, which had a thick, flattened shell 1 metre long. Although *Basilemys* was distantly related to the aquatic softshell turtles, it lived on land like modern tortoises and ate tough, low-growing plants.

More than 10 species of lizard have been identified in the Hell Creek Formation. Most of them, including *Obamadon* and *Chamops*, belonged to an extinct group of herbivorous lizards called polyglyphanodontians. *Cemetarius monstrosus* and *Palaeosaniwa canadensis* were large predatory lizards, *Palaeosaniwa* reaching lengths of 2 metres. In appearance and lifestyle, these lizards may have resembled goannas or gila monsters, stalking the riverbanks and forest floor for prey that included other lizards, mammals and unattended dinosaur eggs and hatchlings. One of the few fossil snakes discovered in the Hell Creek Formation is the 2-metre-long *Cerberophis rex*.

Insects

Because they lack hard body parts, insects are only rarely preserved as fossils, and only in special circumstances. The Hell Creek Formation is no exception. Fossils of insects in the Hell Creek Formation include insect bodies preserved in amber and traces of insect-feeding on fossil leaves. A damselfly, like the ancient greenling that only exists today in southeast Australia, plus several flies related to living blowflies and mosquitoes have been found in amber. Scale insects, leaf-miner flies and leaf beetles have been identified as the culprits of a variety of holes, cuts and tunnels in fossil leaves. Cretaceous fossils from elsewhere in North America, plus evolutionary genetics of living species, show that there was likely to be a multitude of different ants, wasps, bees, moths and butterflies in the Hell Creek ecosystem. This minibeast majority would have added their familiar swarms, nests, sounds and colours to the terrain, air and foliage of long ago.

Baenid turtle

Champsosaurus

Basking on riverbanks alongside turtles were crocodylians and their relatives. The most common of these was an ancient relative of living crocodiles and alligators called *Borealosuchus*, which grew to 4 metres long. Perhaps the strangest reptile of the Hell Creek ecosystem was *Champsosaurus*, which looked a bit like a slender-jawed crocodile but was in fact a choristodere, a group that split from other reptiles more than 250 million years ago. *Champsosaurus* had nostrils at the tip of its long snout, which it may have used like a snorkel while lying in wait to snare small fish in the shallows.

LEFT TO RIGHT

Quetzalcoatlus (pterosaur)

Ankylosaurus (armoured dinosaur)

Palaeosaniwa (lizard)

Pectinodon (troodontid dinosaur)

Edmontosaurus (duck-billed dinosaur)

Trionychid turtle

Acheroraptor (dromaeosaurid dinosaur)

Tyrannosaurus rex

Borealosuchus (crocodile relative)

Pterosaurs

The only pterosaur that we know existed in the Hell Creek ecosystem is a *Quetzalcoatlus*-like azhdarchid. With an estimated wingspan of 5 to 5.5 metres and standing height of 2 to 3 metres, it would have been just half the size of the largest *Quetzalcoatlus* fossils found in Texas! *Quetzalcoatlus* had long toothless jaws, a long stiff neck and long legs. These pterosaurs may have mostly glided while airborne looking for carrion, although their enormous size and ability to walk on all four limbs may have made them more at home on the ground, where they stalked prey along waterways and in open woodland.

Dinosaurs

Tyrannosaurus. Ankylosaurus. Edmontosaurus. Pachycephalosaurus. Triceratops. These names belong to some of the most iconic dinosaurs, and they were the supreme—and perhaps last—examples of their respective groups. They all lived together in the Hell Creek ecosystem. A thorough count of fossils from the Hell Creek Formation calculated that *Triceratops* was the most common dinosaur, followed by *Tyrannosaurus* and *Edmontosaurus*. As we shall see, a diversity of dinosaurs existed alongside the 'Big Five' in the Hell Creek ecosystem.

The bizarre alvarezsaur *Trierarchuncus prairiensis*.

Trierarchuncus prairiensis was a 150-centimetre-long alvarezsaur, which are strange little coelurosaurs. Their extremely short, powerful arms bear a large, hooked claw probably used for ripping apart wood and soil to feed on ants or other insects.

Theropods

Every theropod dinosaur in the world of *Triceratops* was a coelurosaur: that is, a theropod more closely related to birds than to other theropods like the Jurassic *Allosaurus*. The largest coelurosaur in the Hell Creek ecosystem was *Tyrannosaurus rex*, weighing more than 7000 kilograms, and the smallest coelurosaur was a bird weighing about 200 grams. It is likely that every Hell Creek theropod had feathers in some form somewhere on its body—even *Tyrannosaurus* may have had the odd hair-like feather on its head or back.

Tyrannosaurus rex (*T. rex*) was the only medium- to gigantic-sized land predator in the Hell Creek ecosystem. Young *T. rex* were lightly built and faster than fully grown adults and probably preyed on smaller, faster animals. With increasing age and bulk, *T. rex* individuals could begin to tackle slower, larger and more dangerous prey, including *Edmontosaurus* and *Triceratops*. By targeting different prey as they grew up, *T. rex* seems to have left no space in the ecosystem for any other large predatory dinosaurs. *Tyrannosaurus rex* was the only large theropod in the world of *Triceratops!*

Smaller predatory theropods included the sickle-clawed dromaeosaurid *Acheroraptor temertyorum*, a close relative of the infamous *Velociraptor*, and troodontids like *Pectinodon bakkeri*. *Acheroraptor* was about 2 metres long and weighed 18 kilograms, whereas *Pectinodon* may have been 2 metres long with a weight of up to 30 kilograms. Despite having a smaller body than the troodontids, *Acheroraptor* had more heavily built jaws and flesh-slicing teeth that enabled it to target larger prey including non-bird dinosaurs. The smaller, more numerous teeth and narrower jaws of troodontids could mean they preferred smaller prey such as insects, lizards, and mammals.

Other non-bird theropods in the Hell Creek ecosystem were a collection of oddities.

The emu-like ornithomimids had long necks and ran fast on powerful legs with hoof-like toe claws. It's likely that their toothless beaks were mostly used to crop vegetation. The 2.5-metre-tall oviraptorosaur *Anzu* had a short toothless beak, a cassowary-like crest on its head, long legs and long arms with slender claws. This bizarre-looking dinosaur could run fast and had a wide diet of plants and small animals.

Fossils of living dinosaurs (birds) from the Hell Creek Formation show that at least 17 species waded past, flew over and perhaps perched on *Triceratops*. All these birds belonged to extinct groups. The largest was *Avisaurus*, a strong flyer with a wingspan of up to 2 metres. *Avisaurus* belonged to an ancient group called enantiornithines that had scaly jaws lined with sharp teeth instead of a beak. *Avisaurus* had long, sharp claws on its feet for perching and perhaps grasping prey. *Brodavis* was a 40-centimetre-long hesperornithiform that used its enlarged feet to dive underwater in pursuit of small fish that it caught with a toothed beak that was long and narrow.

TOP *Avisaurus archibaldi*

BOTTOM *Brodavis baileyi*

Ornithischians

Most large herbivores in the Hell Creek ecosystem were ornithischian dinosaurs. One of the smallest ornithischians was the bipedal *Thescelosaurus neglectus*, which was 3 metres long and weighed 90 kilograms. *Thescelosaurus* had heavily built arms and legs and was unlikely to have been a fast runner. It may have used its narrow beak to browse from shrubs that grew along riverbanks, which were perhaps its preferred habitat. At the other end of the size spectrum was the huge hadrosaurid (duck-billed dinosaur) *Edmontosaurus annectens*. Most skeletons of this mega-duckbill are between 8 and 12 metres long and represent animals weighing 6000 to 7000 kilograms when alive. But a few *Edmontosaurus* fossils hint that this species grew to even more colossal proportions. The wide, squared-off and bony snout of *Edmontosaurus* was covered by a thick, hooked bill in life. *Edmontosaurus* spent most of its time on all fours, but was capable of rearing up on its back legs. *Edmontosaurus* may have preferred to live in herds closer to rivers.

Rare ornithischians of the Hell Creek Formation include the dome headed *Pachycephalosaurus*, small ceratopsian *Leptoceratops* and the uncommon ceratopsid *Torosaurus*. The wide-hipped *Pachycephalosaurus wyomingensis* grew to 4.5 metres long and 450 kilograms in weight. The teeth of *Pachycephalosaurus* are small and were possibly unsuited for shearing tough vegetation. Instead, *Pachycephalosaurus* may have fed on softer leaves and fruit, or even been an omnivore and included animals in its diet. *Leptoceratops* was a 2-metre-long quadrupedal herbivore that had clawed hands and feet. Its powerful jaws and teeth could be used to shear and crush tough and fibrous low-growing plants in a chewing action more like that of mammals than ceratopsids like *Triceratops*.

The tanks of the Hell Creek ecosystem were the heavily armoured ankylosaurs. Two species have been identified: the nodosaurid *Denversaurus schlessmani* and the ankylosaurid *Ankylosaurus magniventris*.

The 6-metre-long *Denversaurus* had a pointed snout, large spikes on its shoulders and no bony club at the end of its tail. Remains of ferns dominated the fossil stomach contents of another nodosaurid called *Borealopelta*, so ferns and other low-growing plants may have been on the menu for *Denversaurus*. *Ankylosaurus* was the largest of all ankylosaurs, reaching 8 metres long, weighing 5000 to 8000 kilograms and having a tail armed with a large club at its tip. *Ankylosaurus* had a wide, blunt snout that enabled it to graze on ferns and low shrubs.

Two young *Tyrannosaurus* encounter a herd of *Edmontosaurus annectens*.

THE LIFE OF
TRICERATOPS

Triceratops was a mighty quadrupedal animal of massive proportions. Estimates of the maximum size of Triceratops *vary from 6 to 9 metres in total length.*

The problem with these estimates is that there are few skeletons of *Triceratops* that include the skull and a complete vertebral column. Almost all skeletons of *Triceratops* displayed in museums are composites, made up of the real fossils and replica bones from two or more individuals. So how do scientists know how big one singular *Triceratops* could actually get? The few nearly complete skeletons are all between about 6 and 7 metres in length (measured in a straight line). However, those *Triceratops* individuals were probably not fully grown when they died.

Bones of exceptionally large *Triceratops* hint that mature individuals could have exceeded 7 metres in length. Measured to the highest point on the back, large *Triceratops* could be up to 2.5 metres in height. Recent estimates of the body mass (weight) of a living *Triceratops* range from 6080 to 13,274 kilograms.

The head of *Triceratops* was huge. Its weight in a living individual has been estimated to have reached 730 kilograms—the size of a large horse—and its length was about one-quarter to one-third of the entire body of the *Triceratops*. Perhaps surprisingly, the weight of both brow horns in a living *Triceratops* was probably less than 50 kilograms.

A short and strong neck connected the head to a barrel-shaped torso with a balloon-like chest and gut region. The widely flaring ribcage was broader than the hips, with the rear end of the ribcage narrowing abruptly in front of the hips to make room for the long and powerful thighs.

The back was smoothly arched between the neck and base of the tail, which itself was about as long as the torso and was quite narrow along its length. The forelimbs were much shorter than the hindlimbs, and both hand and foot digits ended in robust rounded hooves instead of pointed claws.

In *Triceratops*, like other ceratopsids, the first three neck (cervical) vertebrae are fused into a sturdy block called the syncervical, which together with the six neck vertebrae behind it make a total of nine cervical vertebrae. Between the neck and the hips there are 12 back (dorsal) vertebrae. Ten vertebrae are fused into a block called the sacrum that connects the vertebral column to the pelvis. The number of tail (caudal) vertebrae can be quite different between ceratopsid species, but in Melbourne Museum's *Triceratops horridus* skeleton there are 36 caudal vertebrae.

Under most conditions, skin rots away extremely quickly after an animal dies, so is rarely preserved in the fossil record.

It is no surprise then that after 66 million years, no actual skin tissue has been found with any *Triceratops* fossils. But that's not the end of the story! When some *Triceratops* carcasses came to rest on a surface and were buried, the texture and pattern of their skin left an imprint in the surrounding sediment before the skin decomposed. These copies of the skin hardened to rock and show us what the skin of *Triceratops* looked and felt like.

Did *Triceratops* have feathers, filaments or quills? Fossils from China of the early ceratopsian *Psittacosaurus* have been found with long, bristle-like filaments projecting upwards along the top of the tail like a crest. This at least raises the possibility that *Triceratops* could have had similar wispy filaments or bristles growing somewhere on its body, but there is no fossil evidence that it did. The same exceptional *Psittacosaurus* fossil even gives some hints to skin colouration in *Triceratops*. It preserves skin with its structure and original pigment in a mineralised layer on and around the skeleton. This *Psittacosaurus* had a lightly coloured underside and darker, perhaps brown, skin on its back and sides for camouflage in its forest habitat. If one ceratopsian had skin colour patterns like this, *Triceratops* could have had them too.

Natural cast of the skin of a *Triceratops* fossil showing the large size and shape of the scales.

Scaly skin

Over its hip and thigh region the surface of the skin was formed by a mosaic of slightly raised hexagon-shaped scales about 5 centimetres across, with larger 10-centimetre-wide scales scattered 20 to 30 centimetres apart. Each of the larger feature scales had a short, pointed cone in its centre. On the underside of its neck, *Triceratops* had a flat pavement of rectangular scales of irregular size, most being about 3 centimetres long and 2 centimetres wide. If you could travel back in time and rub your hand across the flank of *Triceratops*, its skin would have felt bumpy. But if you gave *Triceratops* a scratch under its neck, the skin would have felt smooth.

The skeleton of *Triceratops* includes approximately 313 individual bones, 71 of which combine to form a skull that can be 2.2 metres long.

For comparison, there are about 270 separate bones in a human child's skeleton (many of which fuse together in adults).

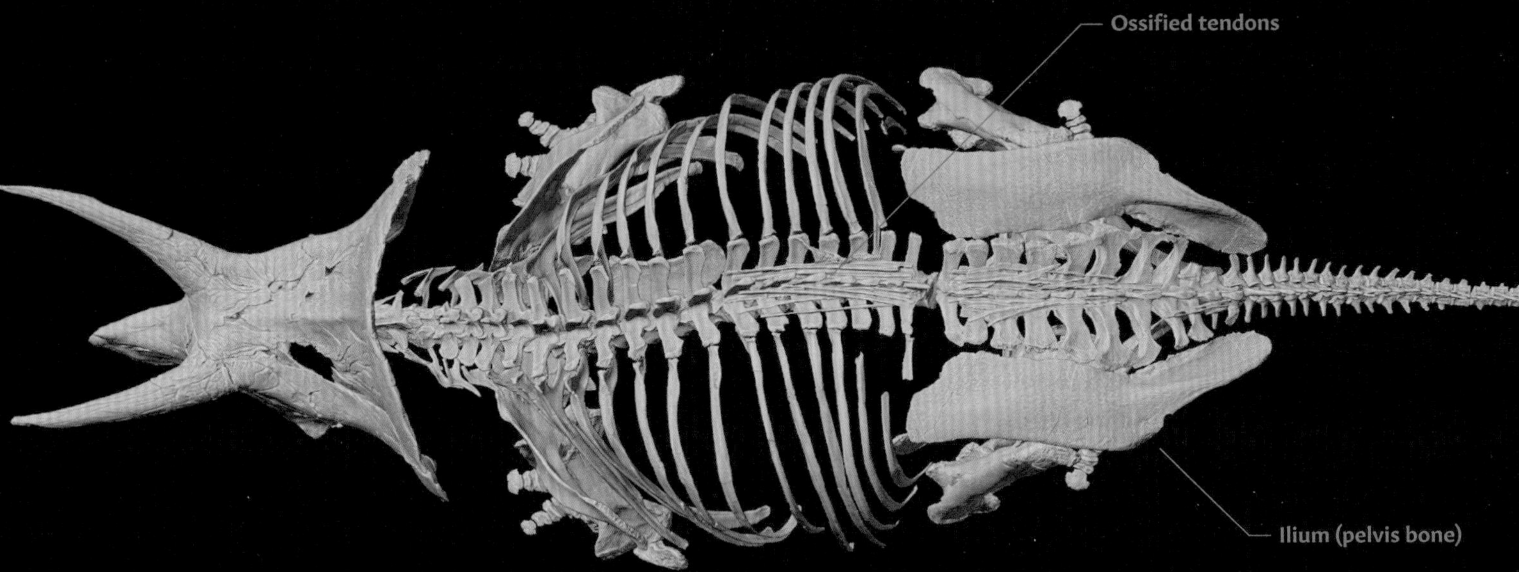

The skeleton of *Triceratops horridus* viewed from above.

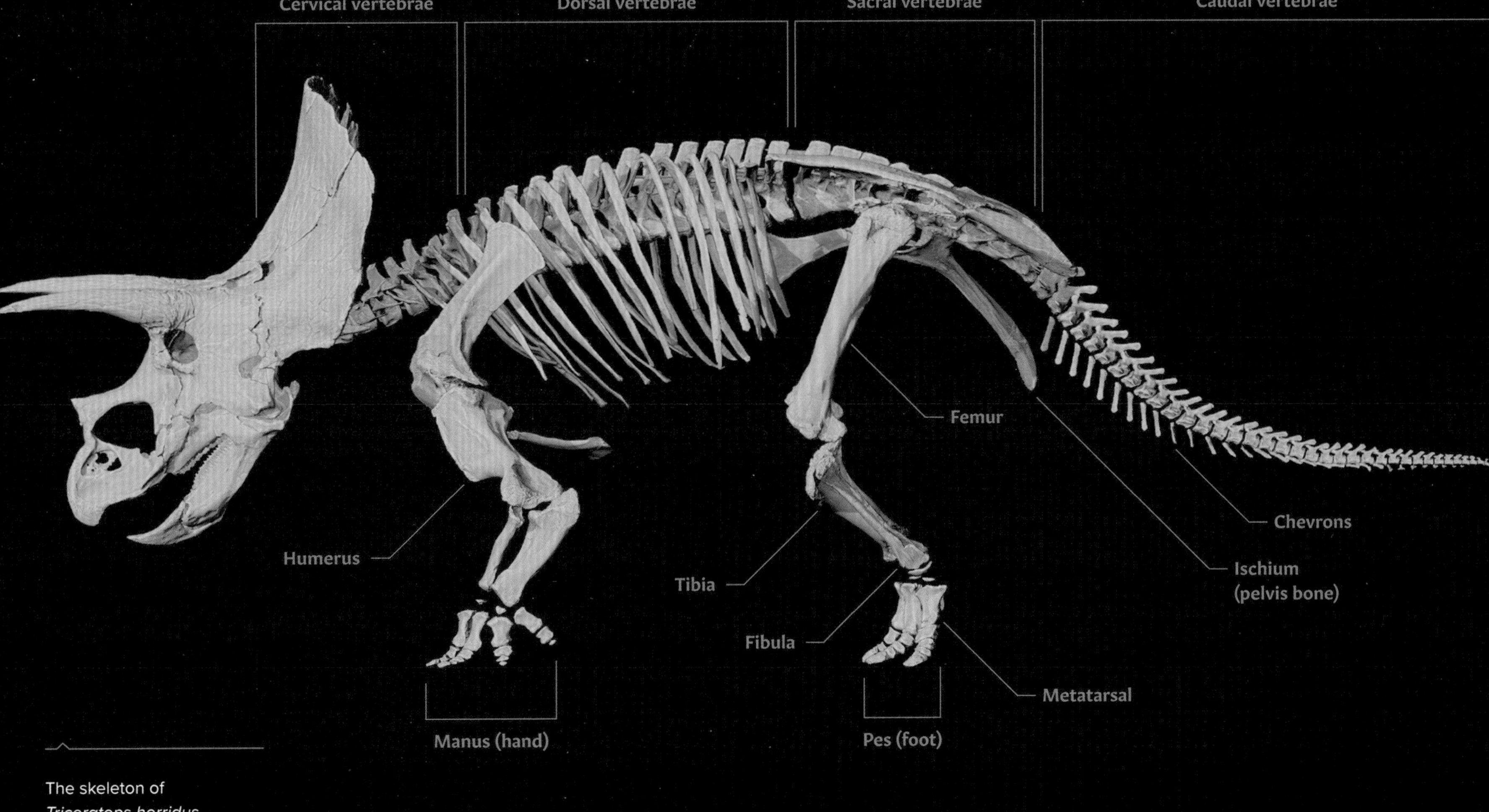

The skeleton of *Triceratops horridus* viewed from its left side.

Could *Triceratops* swim?

Some of the hippopotamus-like proportions of the limb bones in *Triceratops* have been interpreted in other ceratopsids as adaptations to an amphibious lifestyle. *Triceratops* lived in a low-lying coastal environment with swamps and large rivers, so it is highly likely that *Triceratops* had to occasionally wade through water. But venturing into deeper water could have spelled disaster. Computer simulations of the buoyancy of floating chasmosaurines show that their large, heavy heads tip the front half of their bodies forwards, deep underwater. This would have made it impossible to raise their neck and head high enough for the nostrils to break the surface and allow the dinosaur to breathe. Swimming attempts by adult *Triceratops* carried a high risk of drowning.

Posture and Movement

Although its bipedal ancestry is suggested by its hindlimbs being longer than its forelimbs, *Triceratops* were committed quadrupeds that did not habitually stand and walk on their back legs only. Their forelimbs were held in a semi-erect posture, with the elbow bent in a constantly flexed position and pointing slightly out to the side and towards the rear of the body. The hindlimbs were held upright. This strange combination of forelimbs held in mid-push-up and hindlimbs in an upright position has no equivalent in any living animal. This has made it challenging for scientists to understand how *Triceratops* moved and if it was capable of running.

The strength of *Triceratops* limb bones has been calculated to be as strong as those of living rhinos, which can slowly gallop. The limb bone proportions of *Triceratops* are also like those of rhinos and hippos, not more specialised running mammals. However, the anatomy of the elbow joint in *Triceratops* may have stopped it from moving its arm forwards and backwards enough to allow running. No ceratopsid footprint or trackway fossils have been found that show they could run, which is not surprising because most dinosaur fossil trackways show animals walking. Ceratopsid tracks found in Colorado in the United States of America are the right geologic age (Maastrichtian) and size to have perhaps been made by *Triceratops*. Based on the stride length estimated from these tracks, it has been calculated that the ceratopsid track-maker was walking at about 4 kilometres per hour. The question of whether *Triceratops* could run remains unanswered.

Feeding and Food

Triceratops and other horned dinosaurs had a sophisticated feeding system with complex teeth and powerful jaws. Like all ceratopsids, the teeth of *Triceratops* have two roots, and the surface of each crown that meets the opposite tooth is nearly vertical. These teeth are arranged in one long row of between 33 and 40 functional tooth positions, forming a tooth battery with a continuous vertical shearing surface like a giant scissor blade. Two to five teeth are vertically stacked in a column at each tooth position. The number of teeth in each column decreases towards the front and rear of the battery. This means a *Triceratops* can have hundreds of teeth in its jaw at any given time, although only the teeth at the top of each row would be used.

The teeth were continuously replaced through the life of the dinosaur so there were never any gaps in the cutting edge of the tooth battery. A crushing or grinding action like the teeth of hadrosaurs—or living horses and cows—was not possible. This is unusual for a large herbivorous animal. Despite this, the teeth of *Triceratops* were more complex than those of living herbivorous mammals because they were formed from five different types of tissues: enamel, cementum and three types of dentine.

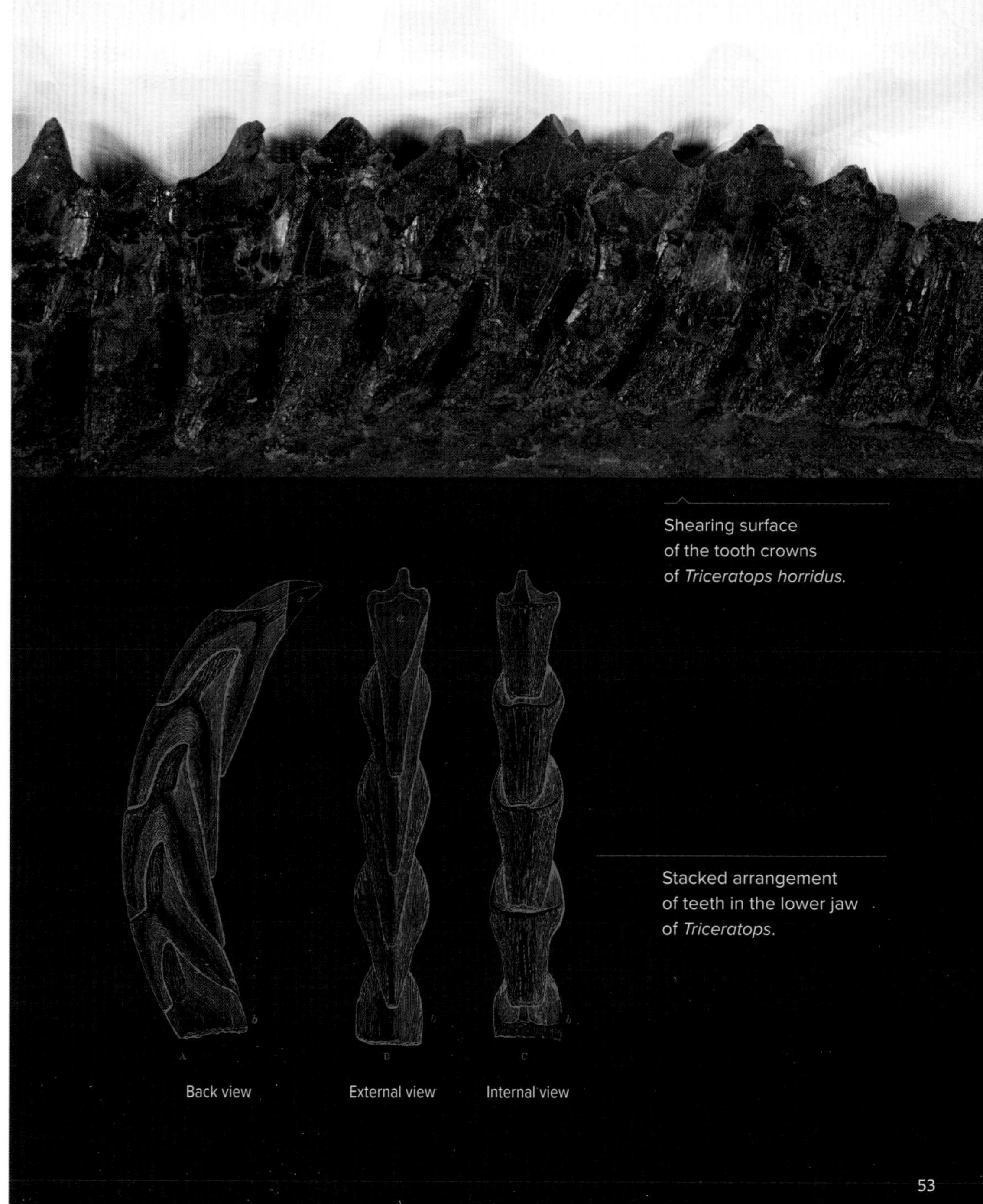

Shearing surface of the tooth crowns of *Triceratops horridus*.

Stacked arrangement of teeth in the lower jaw of *Triceratops*.

What did *Triceratops* eat?

The narrow, pointed upper (rostral) and lower (predentary) beak bones of *Triceratops* were covered in a horny sheath in life, looking much like the hooked beak of a parrot. Its shape suggests it was used to grasp plants like a 'selective feeder', which is an animal that is choosy, preferring to pluck only certain parts of plants instead of consuming anything and everything in front of them.

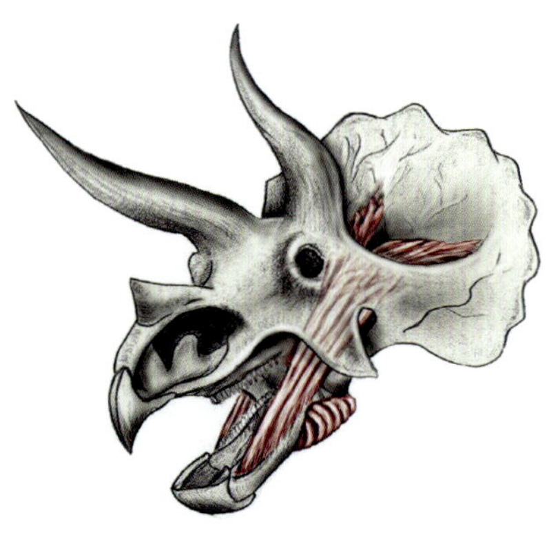

Jaw muscle anatomy of *Triceratops*.

The height of *Triceratops* means that it was probably unable to reach vegetation more than 2 metres above the ground. The fact that *Triceratops* teeth formed shearing blades set in powerful jaws has been interpreted as evidence of a specialised diet of tough fern, cycad or palm fronds. *Triceratops* may have used its formidable head and great bulk to push over small flowering trees, grasp the branches with its beak and then chop up the twigs and leaves with its tooth batteries. Unless the fossilised remains of gut contents are discovered with a skeleton, the diet of *Triceratops* could stay an unsolved case.

The size of the ribcage can give a general idea of how big a dinosaur's digestive tract was. One look at the skeleton of *Triceratops* tells it probably had a lot of guts. It is likely that *Triceratops* was a hindgut fermenter, because most living herbivorous reptiles and birds have this kind of digestive system. In hindgut fermentation, food doesn't spend much time in the stomach, instead being digested in the intestines. This demands a lot of intestines and is very inefficient. A lot of food goes to waste as undigested bits of plants are excreted in the faeces.

Coprolites

Fossilised faeces are called coprolites, and they can provide rare insights into the palaeoecology of extinct animals and their environments. None have been found that are from ceratopsids, but coprolites produced by Late Cretaceous hadrosaurs give some idea of what *Triceratops* faeces could be like. The hadrosaur coprolites range from piles of irregular rounded masses about 14 centimetres thick and 40 centimetres long to more flattened masses with volumes between 5 and 7 litres.

Living elephants can produce 11 litres of faeces in one deposit. One coprolite site preserved multiple fecal deposits with a total volume of 21 litres. The coprolites preserve dung beetle burrows and are full of conifer wood fragments, showing that these dinosaurs were deliberately eating rotting wood. Based on this evidence, piles of *Triceratops* dung may not have been larger than those produced by the largest living land animals.

Horns and Frill

The horns and frill of *Triceratops* are iconic, but their function in the life of *Triceratops* is controversial. Recent research has begun to clear up what were the more likely and less likely roles of these extreme features.

The sizes and shapes of the horns and frill can vary a lot between any two skulls of *Triceratops*. Those differences are often best explained by the skulls being at different stages of growth. Some palaeontologists think that the horns and frills are different because males and females had skull shapes unique to their respective sex. More recent scientific investigations could not find any differences in the bones that could be explained by some *Triceratops* skulls being male and others female.

The brow horn cores of *Triceratops* can be more than 80 centimetres long. The name 'horn core' is for the bone itself. Numerous grooves in the surface of each horn core mark where blood vessels and nerves for a layer of skin covered the bone when the *Triceratops* was alive. This skin formed the base for the non-living outer horn made of the protein keratin, which also makes structures like your fingernails and hair. In life, the keratin horn sheath could have added up to another one third of length to the end of the bony horn core. Picture that next time you look at a *Triceratops* skull! A network of grooves for blood vessels and nerves can also be seen on the front surface of the frill, suggesting it too might have been wrapped in a horny keratin sheath. However, some *Triceratops* fossils have been found with impressions of scales on the frill, so the appearance of the frill in life is still open to debate.

The horns and frill of *Triceratops* were imposing visual signals.

There are three main possibilities for the function of *Triceratops'* brow horns: defence and combat, visual display or a combination of both.

Fossils of *Triceratops* brow horn cores have been found that have broken tips and healed wounds caused by *Tyrannosaurus rex* bites. Palaeontologists think these injuries were caused by Tyrannosaurus bites because the size and shape of the punctures in the *Triceratops* horn core perfectly match the tips of *T. rex* teeth. The fact that the bone eventually healed is strong evidence that the injuries to the *Triceratops* were inflicted while it was alive. The location of the *T. rex* bites on the brow horns suggests these dinosaurs were facing each other when the interaction took place. It is not possible to say if *T. rex* or the *Triceratops* was the attacker in this example. What is clear is that the brow horns of *Triceratops* were positioned pointing towards a predatory dinosaur at close quarters, supporting the idea that the brow horns could have been used in anti-predator defence if needed.

Healed injuries around the base of the frill in several *Triceratops* skulls were probably caused by the impact of the brow horns of rival *Triceratops* during combat. How might clashes between two *Triceratops* have taken place? Perhaps there was an initial frontal threat display with the head tilted forwards, the full length of the frill vertical, and horns pointed towards the opponent.

This show of force could have deterred physical violence if one *Triceratops* was clearly larger than the other. If the opponents were of similar size, combat via head-pushing and wrestling with interlocked horns would have occurred. The large spherical occipital condyle at the back of the skull, which connected the massive head to the neck at a ball and socket joint, allowed a wide range of motion of the head during the duel. These twisting and shoving movements, with horns locked in combat, were powered by huge neck muscles that attached to the back of the frill.

Bring *Triceratops* to life!

Open the *Triceratops: Melbourne Museum* app on your mobile phone or tablet.

If you haven't downloaded the AR app yet, flip back to the contents page at the beginning of the book and scan the QR code to get started.

Follow the instructions in the app to scan the *Triceratops* skull on the right and see how clashes between two rival *Triceratops* may have taken place.

Anatomy of the skull of *Triceratops horridus.*

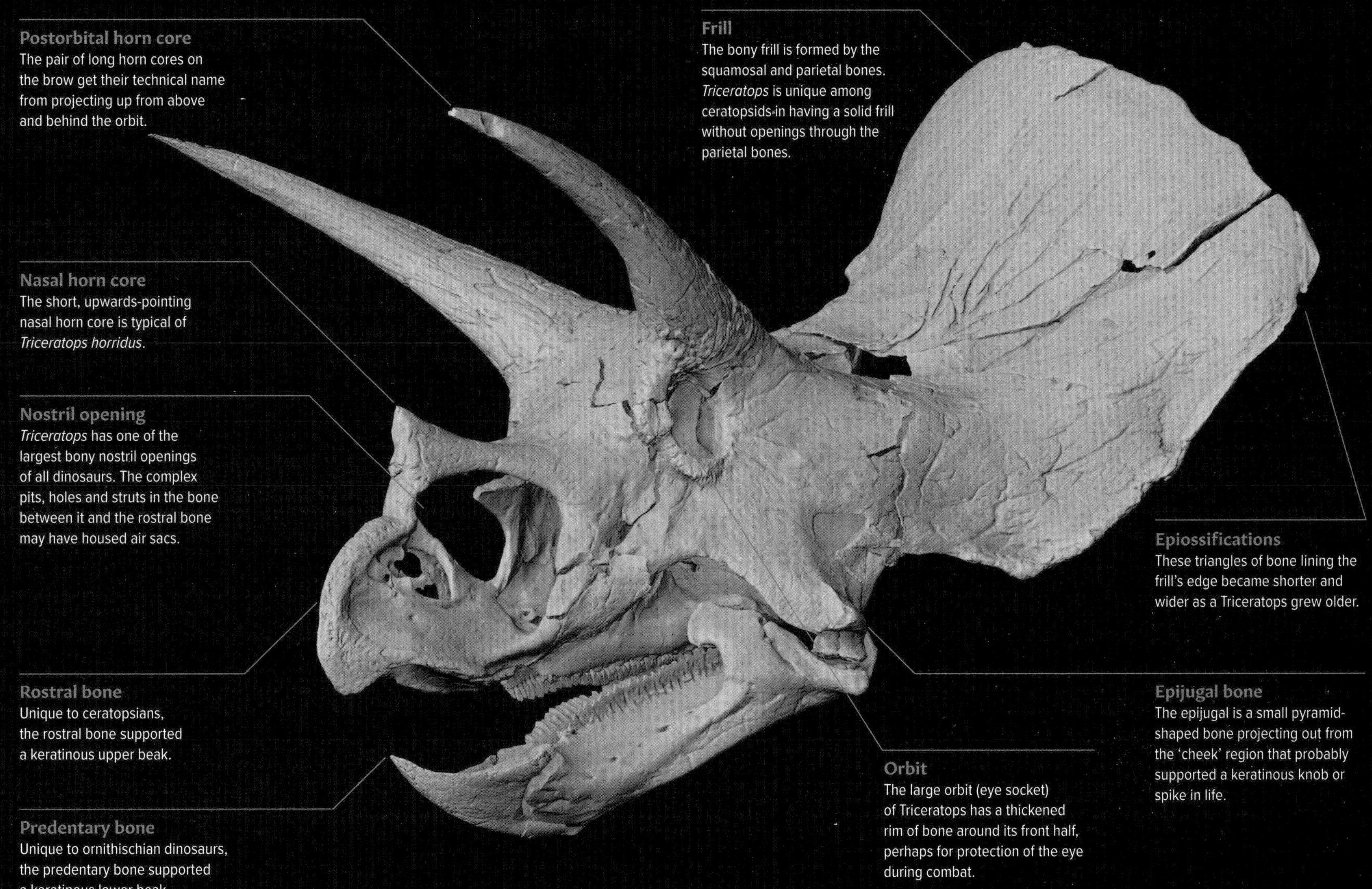
Postorbital horn core
The pair of long horn cores on the brow get their technical name from projecting up from above and behind the orbit.
Frill
The bony frill is formed by the squamosal and parietal bones. *Triceratops* is unique among ceratopsids in having a solid frill without openings through the parietal bones.
Nasal horn core
The short, upwards-pointing nasal horn core is typical of *Triceratops horridus*.
Nostril opening
Triceratops has one of the largest bony nostril openings of all dinosaurs. The complex pits, holes and struts in the bone between it and the rostral bone may have housed air sacs.
Epiossifications
These triangles of bone lining the frill's edge became shorter and wider as a Triceratops grew older.
Rostral bone
Unique to ceratopsians, the rostral bone supported a keratinous upper beak.
Epijugal bone
The epijugal is a small pyramid-shaped bone projecting out from the 'cheek' region that probably supported a keratinous knob or spike in life.
Orbit
The large orbit (eye socket) of Triceratops has a thickened rim of bone around its front half, perhaps for protection of the eye during combat.
Predentary bone
Unique to ornithischian dinosaurs, the predentary bone supported a keratinous lower beak.

Was *Tyrannosaurus rex* a predatory threat to *Triceratops?*

Were these dinosaur titans eternal enemies, forever facing-off as often depicted in popular culture? We have seen that there is some direct fossil evidence of violent encounters between these iconic species.

The fact that the *T. rex* bites on the *Triceratops* horn healed shows that the *Triceratops* survived. If the scenario was an attack by a *T. rex*, it was an unsuccessful predation attempt. The reality is that an adult *Triceratops* would have been a formidable and difficult animal for *T. rex* to tackle compared to other potential prey. So predatory attacks on large adult *Triceratops* may have been infrequent and only attempted by the largest, or most inexperienced, *T. rex* individuals. *Tyrannosaurus* could have preferred preying on younger, smaller *Triceratops* that were lower-risk options.

The use of the brow horns in fighting pairs of *Triceratops* might suggest that the main function of the frill was a shield to protect the neck.

Compared to all other ceratopsids, the frill of *Triceratops* was unusual because it was thickened, solid bone rather than thin bone with large openings in it. This tall sheet of bone projecting up and backwards from behind the eye sockets would have required a lot of energy to grow. So whatever it was doing, it must have been vitally important to *Triceratops*' biology.

As ceratopsids grew from baby to adult, the frill changed in shape and size faster than any other part of the skull, only reaching its greatest development when the animal was fully grown. This is evidence that the frill was a visual signal to other *Triceratops* in social and reproductive behaviour, much like the elaborate feathers and courtship displays of peacocks and birds of paradise. Because all ceratopsids besides *Triceratops* have a frill formed of thin bone with large openings in it, it seems the evolution of the frill was driven mainly by its visual display role. So, the frill of *Triceratops* is a billboard first and armour second. Why no other ceratopsids with long brow horns evolved a solid, thick frill like *Triceratops* is a mystery.

OPPOSITE A flock of ornithomimid dinosaurs pass by an aggressive encounter between two *Triceratops*.

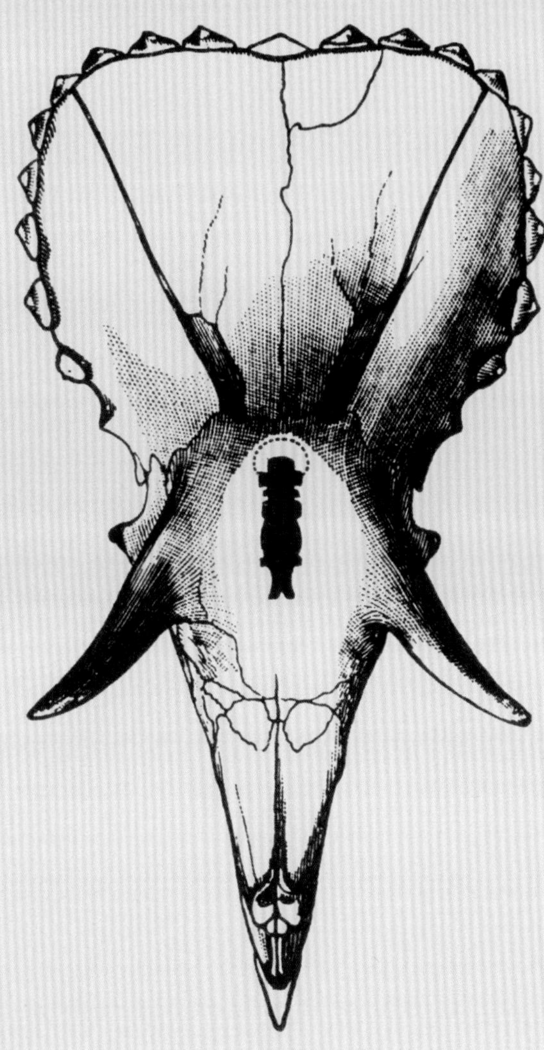

The brain of *Triceratops* (shaded black) was about the size of an orange, but in a skull the size of a horse.

Brain and Senses

We can get an idea of the size and structure of the brain of *Triceratops* by studying casts of the cavity inside the skull (endocasts) that housed the brain. Endocasts of the *Triceratops* brain cavity have a volume of 300 to 400 centimetres^3. For comparison, the volume of a human brain can exceed 1100 centimetres^3. So the absolute size of *Triceratops*' brain was modest. If the brain size of *Triceratops* is compared against body weight, *Triceratops* had a brain slightly smaller than expected for a modern crocodile if the crocodile were the size of a *Triceratops*. That puts *Triceratops* ahead of the long-necked sauropods, ankylosaurs and stegosaurs in the relative brain-size stakes!

Smell

The proportions of different parts of the brain can also tell us something about the capabilities of the senses in an animal. An endocast from one *Triceratops* suggests that the part of the brain involved with the sense of smell, the olfactory bulbs, was small compared to endocasts of similar-sized dinosaurs or smaller, more ancient ceratopsians. This could mean that sense of smell was not hugely important to the lifestyle of *Triceratops*. If that is the case, it makes the enormous bony nostrils of *Triceratops* even more unusual. Why have such large nostrils, plus complicated structures in the surrounding bones, if not for super-smelling? It seems that the nostril area of *Triceratops* was plumbed with air sacs connected to air-filled spaces (sinuses) inside the upper jaw bones. The function of these nasal air sacs remains a mystery.

Hearing

The inner ear (cochlea) of *Triceratops* was tuned to hear low frequency sounds, being most effective at 290 hertz but capable of hearing up to about 1500 hertz. This is within the range of hearing of both chickens and humans, although we can hear much higher frequencies than those dinosaurs. Low frequency sounds don't scatter much over long distances or through dense habitats, so *Triceratops* may have been able to hear sounds from far away in its thickly vegetated environment. Those sounds could have been the vocalisations of another *Triceratops*, but they could also have been the low frequency sounds made by *Tyrannosaurus*, which was capable of hearing—and perhaps producing—a similar range of frequencies to *Triceratops*.

Sight

The living animals most closely related to *Triceratops*, crocodiles and birds, have colour vision, so it is reasonable to say that *Triceratops* also saw the world in colour. So, were *Triceratops* eagle-eyed too? Substantial eye-size has been linked to acute vision, and the diameter of a large *Triceratops*' eye socket can reach 13 centimetres. That's big enough to hold a grapefruit. That doesn't necessarily mean the eyeballs were the size of grapefruits, because the eyeballs of living animals don't always fill the entire eye socket, but the eyes were certainly large. Research on the eye sockets of dinosaurs has shown that nearly all ornithischian relatives of *Triceratops* had eye sockets most like living animals active during the daytime and twilight. This evidence suggests *Triceratops* was unlikely to be nocturnal.

Social Behaviour

It is tempting to think that *Triceratops* lived in herds, based on both the abundance of its fossils and the image of grasslands teeming with herds of bison, buffalo and antelopes—horned mammals that superficially recall the appearance of *Triceratops*. Unfortunately, there is no compelling fossil evidence to support this vision. Other species of ceratopsid, most notoriously *Centrosaurus*, have been found in their hundreds in fossil bonebeds resulting from catastrophe when large numbers of the same species got caught in a flood at the same time. These bonebeds suggest that at least some horned dinosaur species did live in herds. It may be that adult *Triceratops* individuals spent much of their time living alone, coming together in small groups in places where favourite food was abundant or when it was time to breed.

Reproduction and Growth

How did *Triceratops* mate? A fossil of the ceratopsian *Psittacosaurus* preserves the mineralised remains of a cloaca, the all-purpose opening for defecation, urination and reproduction. This is strong evidence that other ceratopsians, including *Triceratops*, would have had a cloaca, which is not surprising because the males and females of living birds and reptiles have one too. Copulation with a cloaca requires a male to press the outer opening, or vent, of its cloaca to that of a female. The cloacal vent is located on the underside of the tail, just behind the legs. Mating pairs of multi-tonne, long-tailed *Triceratops* must have carefully manoeuvred this region of their bodies into close contact to reproduce. If living reptiles and birds are anything to go by, *Triceratops* copulation would have been brief compared to potentially long and elaborate courtship rituals.

Based on evidence from *Protoceratops*, *Triceratops* may have laid soft-shelled eggs that were symmetrical and ellipsoid in shape. Fossilised eggshell of ornithischians lacks pigment, so *Triceratops* eggs were most likely uniformly white or cream-coloured. Although we have no direct evidence from *Triceratops* or any other ceratopsid, fossils of *Protoceratops* hint that *Triceratops* could have laid clutches of up to 15 eggs on or in the ground.

Adult Triceratops *would have been far too large to incubate their eggs by sitting on them—imagine an adult elephant sitting on a pile of chicken eggs, and one can see this strategy ending in extinction.*

Like living crocodiles, they probably either built nests out of mounds of plant material or buried their eggs in moist soil, relying on heat from decomposing vegetation to incubate them. Recent research counted daily growth rings in the teeth of *Protoceratops* and hadrosaur embryos in fossil eggs, showing they took between two and six months to incubate. If *Triceratops* eggs also incubated slowly over several months, they would have been at risk of predation or destruction by environmental events like drought, flood or fire. Could *Triceratops* parents have guarded the nest while the eggs were incubating? No known fossils of *Triceratops* give an answer, but a fossil nest of young *Protoceratops* might offer a clue. The 15 little *Protoceratops* skeletons found in the nest were clearly very young, but not hatchlings—their bones show that they had stayed in the nest, survived and grown for some time after hatching. So perhaps these young *Protoceratops* were being cared for by a parent, and *Triceratops* might have also shown this behaviour before and after their eggs hatched.

Triceratops may have had its iconic three horns and frill as soon as it hatched.

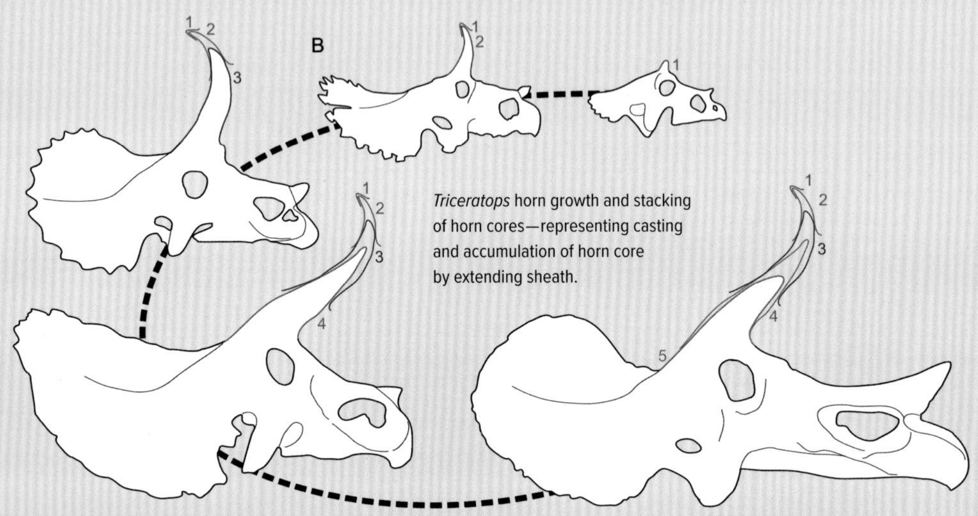

Triceratops skull and horn growth from baby to adult.

By comparing a set of tiny to huge *Triceratops* skulls, palaeontologists have shown that *Triceratops* skulls changed dramatically in shape during growth from baby to adult. The smallest known skull had a short frill and brow horn cores just 3.5 centimetres long, but it is unknown if they were born with these stubby horns. As the skull grew larger and larger, the brow horn cores got longer and their tips gradually rotated from pointing up and backwards in juveniles to pointing up and forwards in large adults.

One baby *Triceratops* skull has been found. At just 30 centimetres long, this skull is tiny compared to the largest adults, but it is larger than we'd expect for a hatchling *Triceratops*.

Nestling *Protoceratops* skulls are fewer than 5 centimetres long, which gives an idea of just how small hatchling *Triceratops* may have been. When juvenile and small *Triceratops* left the nest they could have formed mixed-age social groups for safety in numbers, as suggested by clusters of young *Psittacosaurus* and *Protoceratops* skeletons.

Microscopic structures inside bones—even the shapes and patterns of bone cells—get preserved in fossils, allowing palaeontologists to study how the bones changed between young and older dinosaurs. If there are tiny lines of growth within the bones, they can be counted like tree rings to get an estimate of the age of the animal when it died. This research has shown that most non-bird dinosaurs grew up very fast. Scientists have not yet calculated the age (in years) of any *Triceratops* fossils, but have had success with other ceratopsians. *Psittacosaurus* reached full size when it was 15 years old, but it is much smaller in body size than *Triceratops*. What about ceratopsids closer in size and family relation to *Triceratops*? The centrosaurines *Centrosaurus*, *Einiosaurus* and *Pachyrhinosaurus* all grew fastest in the first three to five years of life and reached their reproductive age before they were 10 years old.

Fossil bones sometimes preserve structures that are not normal anatomy and can be evidence of injuries and illness that ceratopsids, including *Triceratops*, must have experienced during their lives. One *Triceratops*' shoulder blade has a large spine of bone that grew inwards towards the ribs and may have made movement of the shoulder difficult and painful. Other ceratopsids show signs of injuries and infections in their leg and hand bones, which they lived with for years but could have caused them to limp when walking. A large lump on a *Centrosaurus* shinbone has been identified as a tumour caused by a cancer, which would have been fatal if the dinosaur had survived the flood that was the likely cause of its death.

How long do *Triceratops* live?

Given their fast growth in a world of many dangers, how long could a *Triceratops* live for? We do not know. The oldest known *Pachyrhinosaurus* died at the age of 19 years, but the internal structure of its bones show that it was still growing.

A fossil of the chasmosaurine *Spiclypeus shipporum* was about 10 years old and fully grown when it died. Looking outside ceratopsians, the oldest known *Tyrannosaurus rex* died at 28 or 29 years of age.

Extinction

Triceratops and every non-bird dinosaur alive at the same time as *Triceratops*, went extinct 66.05 million years ago. This date in geologic time is one of the great turning points in Earth's history. It marks the end of the Cretaceous Period, the end of the entire Mesozoic Era and the beginning of a new one—the Cenozoic, which continues to the present day. The first Period of the Cenozoic Era is called the Paleogene. So, the mass extinction at the boundary between the Cretaceous and the Paleogene is called the Cretaceous-Paleogene, or K-Pg, extinction.

The K-Pg mass extinction was triggered by the impact of a 12-km-diameter asteroid traveling at hypersonic speed, smashing into the ocean at the edge of the present-day Gulf of Mexico.

This almost unimaginable collision caused devastating tsunamis that swept inland across North America. The Earth's crust was vaporized at the site of impact and super-heated gas, rock, dust, and ash blasted into the atmosphere, sparking catastrophic global forest fires, and blocking out the sun. The resulting darkness, global cooling and acid rain led to mass die-off of plants, so food chains and entire ecosystems collapsed.

There is little doubt that the K-Pg event sealed the fate of the non-bird dinosaurs. We may never know exactly what made the non-bird dinosaurs more vulnerable to extinction than other animals, but it was probably a combination of factors. What continues to be debated by scientists is how dinosaur biodiversity was trending *before* the mass extinction, in the last 10 million or so years of the Late Cretaceous. This boils down to two alternative patterns. Either dinosaurs were thriving right up until their sudden extinction 66 million years ago; or they were in gradual decline starting at about 76 million years ago. Research is ongoing, but there does seem to have been a reduction of species in some dinosaur groups, especially ceratopsians, between 76 and 66 million years ago.

Zooming in on the Hell Creek Formation, there is no evidence of a decline in its dinosaur species richness or ecological diversity during the last 2 million years of the Late Cretaceous in Montana. So, was *Triceratops* really one of the last of all non-bird dinosaurs, at least in North America? As we have seen, *Triceratops* fossils from rocks near the top of the Hell Creek Formation have all been identified as the species *Triceratops prorsus*. One brow horn core of a large ceratopsid has been found just 13 cm below the level of the K-Pg boundary in Montana. This is the most geologically recent non-bird dinosaur fossil that has been found anywhere. It shows that non-bird dinosaurs, particularly horned dinosaurs, were still alive on Earth just before the asteroid struck. We know that *Triceratops* was by far the most abundant ceratopsid in the Hell Creek Formation. So, let's speculate that this lone horn core likely once belonged to a *Triceratops*. In that case, *Triceratops* was the last non-bird dinosaur, and it may very well have been eyewitness to the catastrophic events that ended its world.

With the passing of the Cretaceous Period and dawn of the Cenozoic Era, the ecosystem spaces left vacant by non-bird dinosaurs began to be filled by the survivors.

The theropod dinosaurs consolidated their evolutionary explosion, today outnumbering mammal species and entering lifestyles unheard of in the Mesozoic, such as the polar seafaring penguins. The birds living with us now are what that group of dinosaurs, which first appeared in the Jurassic, did with an extra 66 million years of evolution.

OPPOSITE A surviving mammal surveys the world just after the end of the Cretaceous Period.

ONE SPECIAL

TRICERATOPS

Sixty-six million years ago, *Triceratops* and its Late Cretaceous world came to a sudden end. The clay, mud and sand entombing dinosaur bones hardened into sedimentary rock.

Over the next 10 million years, tectonic forces continued to build the Rocky Mountains and the Late Cretaceous rocks rose higher and higher above sea level.

Millions more years of erosion cut down through the sedimentary layers until the rocks of the Hell Creek Formation in Montana were exposed to the elements. The soft sandstone and mudstone easily eroded to form badlands. This natural erosion frequently exposes long-buried dinosaur fossils at the surface. On most days of searching, these clues to the past turn out to be just fragments—maybe one bone. But on some days, just very rarely, these fossils turn out to be something much more special.

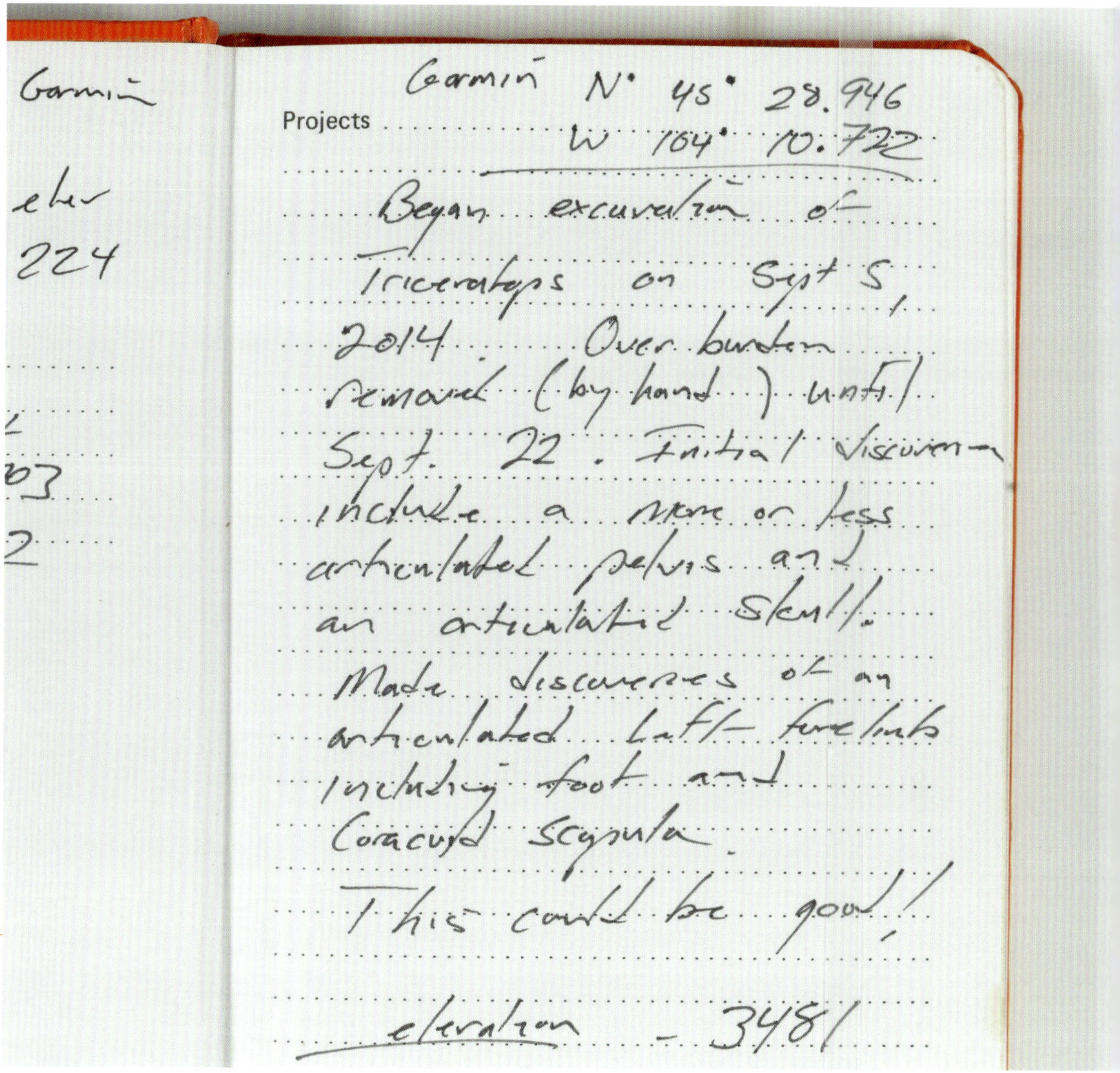

Craig Pfister documented the discovery of the *Triceratops* in a field notebook.

Finding *Triceratops*

One of those days was the 30th of August 2014. Commercial palaeontologist Craig Pfister was on a ranch in south-eastern Montana to finish digging up a *Tyrannosaurus rex* skeleton. On the night of the 29th of August it had rained heavily, making the hillsides slippery and unstable. Instead of setting out in the morning on the quickest route to the *T. rex* site, Craig waited for the hills to dry out. But by mid-afternoon they were still wet. Deciding to forge ahead with the climb up a slope, Craig made his way slowly, cutting steps into the hillside to avoid sliding back down the hill. As he went, he found he had more time than usual to look closely at the outcrop. So it was that in the late afternoon, Craig glanced down into a deep hole eroded out of the slope. The sun was shining down at just the right angle, lighting up the bottom of the hollow. There, poking out of the grey sandstone, was a dark brown object Craig recognised in an instant: the back end of a *Triceratops* pelvis.

Craig went back to camp, gathered a few tools and glue, and returned to the hole in the ground. The hole was at least 3 metres deep, so Craig carefully slid down into the pit. Looking closer at the bone that first caught his attention confirmed it was part of a *Triceratops* pelvis. As he began to glue some cracks in the surface of the pelvis Craig noticed more bone in the wall of the pit. It was the right thighbone (femur), and it was connected, or articulated, with the pelvis almost exactly as it would have been when the animal was alive.

Articulated dinosaur skeletons in the Hell Creek Formation are rare, so even though he could only see two bones it was a sign that this fossil was worthy of more digging.

It was getting late in the season. Craig knew he would need to bring in heavy equipment to finish digging up the *T. rex*. But September in Montana brings snow, and it is not the time to start a major excavation. So, Craig decided to postpone work on the *T. rex* and instead see if there was more to the *Triceratops*.

Over the next three weeks, using only picks and shovels, Craig's field crew dug down through about 3 metres of rock to reach the layer entombing the *Triceratops*. Carefully digging forwards from the exposed rear end of the pelvis, the top of a vertebra appeared. Then another in front of it, followed by another vertebra. An intact, articulated line of vertebrae was soon surfacing above the sandstone. Eventually, at the end of the vertebral column opposite to the pelvis, a broad shield-like frill appeared, and further forwards, two solid horn cores. The skull was there, and it looked like it was complete. Craig now knew this Triceratops fossil was special.

The left femur and tail vertebrae can be clearly seen emerging from the sandstone.

OPPOSITE Articulated tail vertebrae still embedded in sandstone.

The intact skull of the *Triceratops* as it was uncovered in the field.

When the field season ended on the 31st of October, Craig and his team had excavated the entire skull, left forelimb, many ribs and the left tibia—all by hand. By then it was clear Craig had discovered an exceptional *Triceratops*.

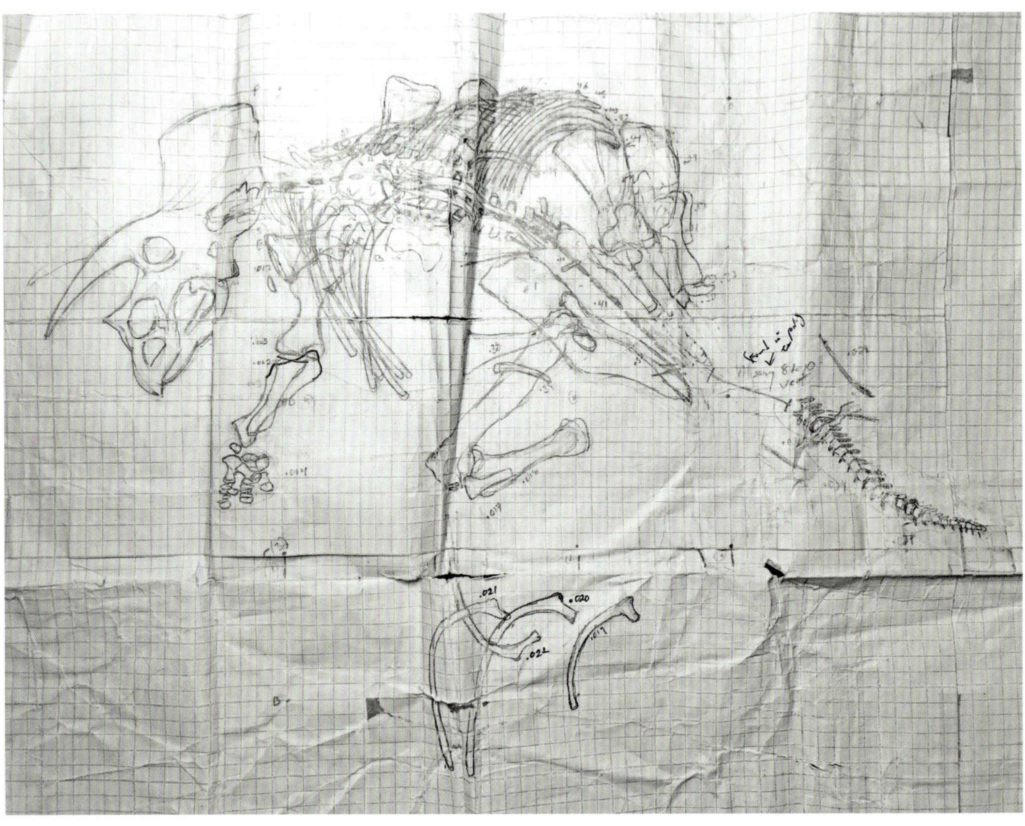

The original field map of the articulated *Triceratops* skeleton, drawn by Craig Pfister.

Craig began to draw a map of the dig site, carefully plotting the position of bones in the ground as they were found—vital for understanding the positions of different bones in the skeleton and how the *Triceratops* was preserved. The huge skull, still partly embedded in sandstone, needed to be lashed on to a sled and pulled out of the dig site with a tractor by the rancher and his family.

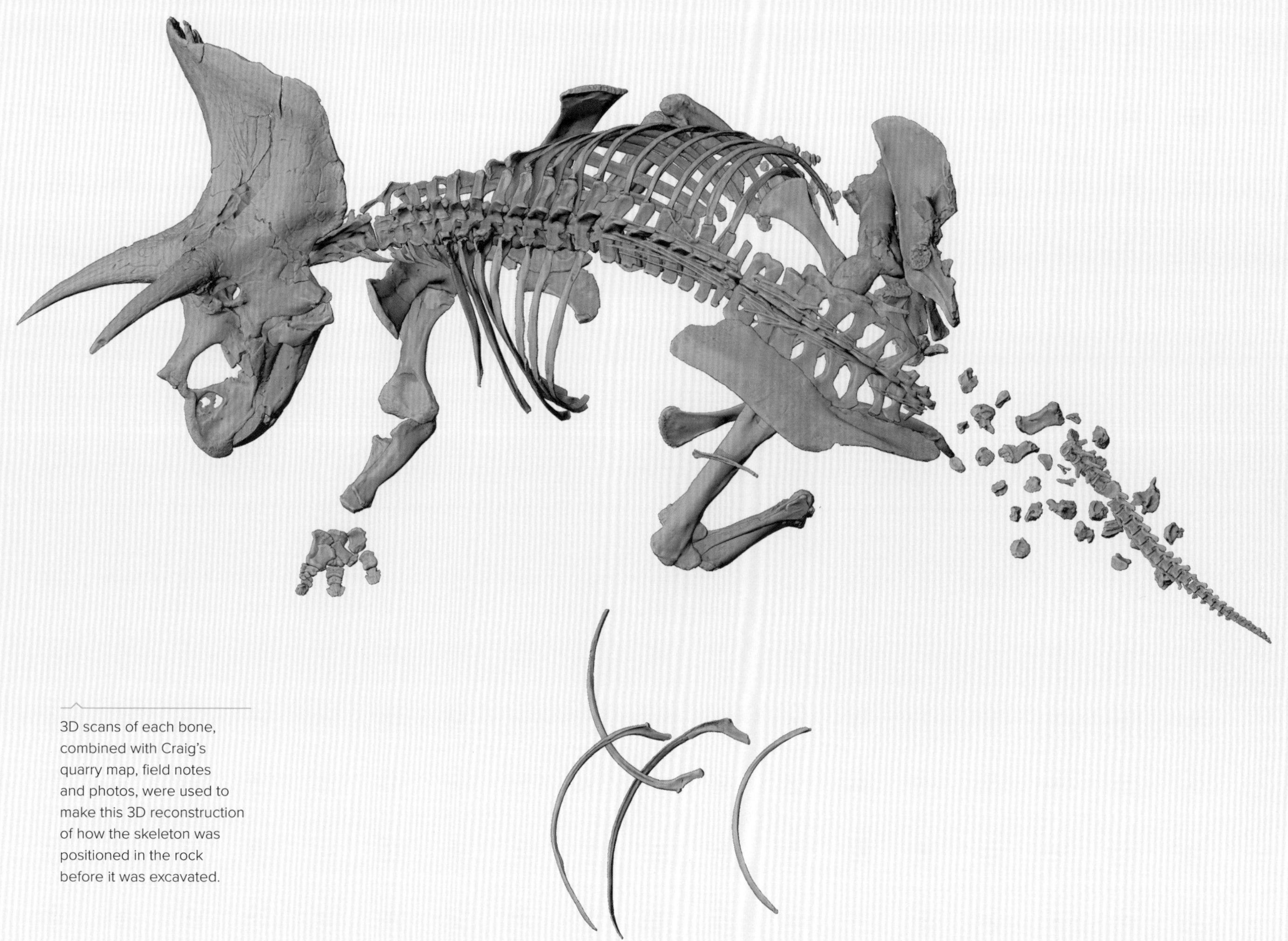

3D scans of each bone, combined with Craig's quarry map, field notes and photos, were used to make this 3D reconstruction of how the skeleton was positioned in the rock before it was excavated.

The second season of the *Triceratops* dig began in June 2015. Excavation required heavy digging machinery to remove more of the bulk rock surrounding the skeleton, followed by more precise close-quarters digging with hand tools. This was possible because the fine sandstone enclosing the fossil is surprisingly soft. The downside to this was that the fossil bones could easily break apart as they were dug out of the sediment.

To reduce this risk, Craig removed different parts of the skeleton in the largest blocks possible, keeping some of the rock around the bones (matrix) so that it would hold together better. Matrix is what palaeontologists call the sedimentary rock in which fossils are embedded.

To make sure these blocks of stone and bone were protected on their journey from field to laboratory, they were each encased in a plaster field jacket.

These were made by carefully wrapping each block in aluminium foil or wet sheets of paper to separate the fossils from strips of burlap (hessian) dipped in wet plaster, which were then wrapped and pressed tightly against the block. For large blocks, several layers of wet plaster-soaked burlap were applied for extra strength. Once dried, the plaster jackets provided solid armour for their precious cargo. By the end of the second season on the 7th of October 2015, about 30 plaster jackets containing the entire *Triceratops* skeleton had been collected. Loaded into the back of heavy-duty trucks, the *Triceratops* fossil was hauled over prairie meadows and dirt roads until it left the ranch, cruising the interstate highway to Craig's workshop in Wisconsin.

Almost three years to the day later, I was standing in the office of Terry Ciotka at his fossil restoration company Dino Lab in Victoria, British Columbia, Canada. Terry had just shown me some photographs of an intriguing *Triceratops* fossil discovered in Montana—the very same fossil discovered and expertly collected by Craig Pfister! I had been sent to North America as a representative of Museums Victoria on a fact-finding mission: if Melbourne Museum were to display a large fossil dinosaur skeleton, how could we go about finding one, and what dinosaur species might be available?

After repeatedly peering at the few photographs of the Montana *Triceratops* and comparing it with other *Triceratops* fossils, I soon realised what Craig Pfister already knew. This was an unusually complete *Triceratops* fossil, with potentially great scientific, educational and entertainment value if it were in a public museum.

I needed to see it for myself to be sure. I got my first chance in July 2019 when I returned to Dino Lab in Canada to look at the fossil in person. After three days of examining as much of the fossil as possible (most of the skeleton was still embedded in sandstone and wrapped in plaster jackets), I was convinced that the *Triceratops* was as good or even better than it looked in the photographs. I reported my findings back to Melbourne. Six months later, Museums Victoria became the permanent home of one of the most complete and best-preserved skeletons of a dinosaurian icon. Horridus, the name given to this remarkable *Triceratops*, was going to call Melbourne home. But first, there was work to do.

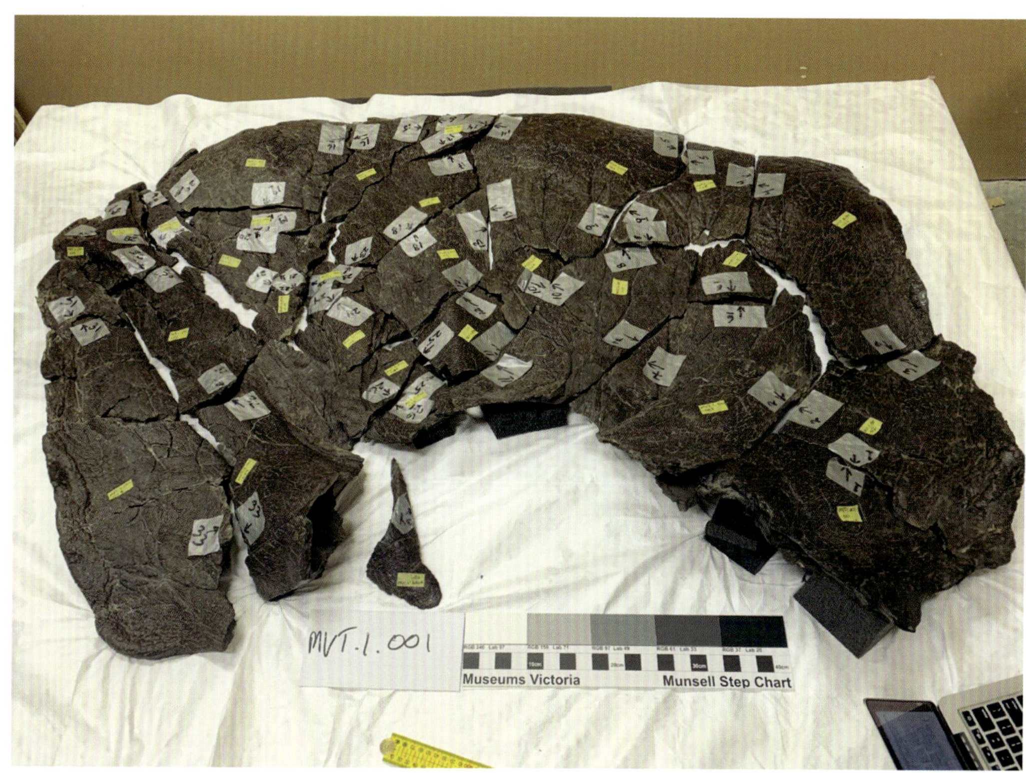

Reassembling the pieces of the frill was like a giant jigsaw puzzle.

OPPOSITE The articulated left manus (hand) in half of its field plaster jacket.

Preparing *Triceratops*

The *Triceratops* skeleton was going to become part of the Palaeontology Collection of Museums Victoria. It would be made available for scientific research and be placed on permanent public display in a major new exhibit at Melbourne Museum. Before any of this could happen, the bones would need to be skilfully taken out of the rock and cleaned in a laboratory. The brittle fossil bones would then need to be hardened, broken parts repaired and missing pieces of the skeleton replaced with artificial copies. This painstaking process is called fossil preparation, and it is the most important step in palaeontology. This is because the planning and care taken in fossil preparation determines the future condition of a fossil, and what can be done with it.

Small hand tools were used for close-up preparation of the *Triceratops*.

The experienced team at Dino Lab got to work preparing the fossil in February 2020. It would take 15 months to fully prepare the skeleton.

Plaster jackets first had to be cut open with a small electric saw. Then specially modified pneumatic engraving pens (essentially miniature jackhammers) were used to chip away the harder sandstone in bulk. Closer to the bones, small hand tools like hobby knives, scraping tools and brushes were used to gently clear the matrix away from the fossil bones.

For a final clean of the surface area of some bones, an air abrasive machine was used to spray a fine powder of sodium bicarbonate at high pressure. As the bone was freshly exposed, a thinned glue solution called a 'consolidant' was applied, seeping into the brittle bone in order to harden it. Broken bones were repaired with specialist glues strong enough to form a steel-like bond between two boulders of granite. The glues, and techniques for putting them on the fossils, were all rated as archival (meaning they don't easily deteriorate) and reversible in case the repair needed to be adjusted.

TOP Larger plaster jackets were cut open with a saw.

BOTTOM The right forelimb bones emerging during preparation. The palm of the right hand (manus) is facing you.

Like these left finger bones, every *Triceratops* bone was assigned a unique catalogue number to track it from preparation to arrival in the Museum collection.

3D surface scanning the vertebrae of the *Triceratops*.

Each bone in the skeleton was given a unique catalogue number, measured, photographed and 3D-scanned with a hand-held surface scanner. This helped keep track of which bone was which, which plaster jacket each bone came out of, and eventually, how each bone fit together in the skeleton. This thorough documentation would prove essential to completing work on the *Triceratops* during the COVID-19 pandemic. It also allowed the museum team in Melbourne to work out which bones were not preserved and would need to have artificial 'stand-ins' made to complete the skeleton for display.

The left and right halves of a dinosaur skeleton are symmetrical, so if we had a bone preserved on one side of the skeleton but missing from the other, the sculpted bone would be a mirror image of the real fossil bone. If a type of bone was not preserved in the skeleton at all, the sculpted replica was based on the anatomy of that bone in other examples of *Triceratops*, or another species of ceratopsid. These replica whole bones, plus some missing pieces of bones, were made from an extremely strong, long-lasting cement and fibreglass sculpting putty. After several days of drying, the attached replica bone parts were strong enough to hold the weight of even the largest limb bones.

All replica parts of the skeleton were painted grey so that it would be obvious which bones are real fossil and which are not. This choice emphasises how little of the *Triceratops* skeleton is not original fossil. It also prevents any confusion over what parts of the fossil are real anatomy when they are being examined by palaeontologists and other museum staff.

Posing *Triceratops*

While the fossil bones were being prepared, repaired and restored, the team in Melbourne was busy planning what the skeleton would look like on display in the museum. First, we had to choose a pose, or look, for the skeleton. Next, the steel support for the bones, called the armature, would need to be designed and made. This is where the theoretical design of the skeleton's pose must sometimes compromise with the physical limits of shapes steel can be bent in, plus take into account the safe placement of heavy, fragile fossils on that steel. The finished display structure, with all the bones cradled on the armature, is called the mount.

The pose of the skeleton mount had to be scientifically credible, make the fossil look 'alive' and emphasise the unique qualities of this Triceratops.

It also had to be a pose for which an armature could be built that would safely hold up the fossil. But the decisive factor in selecting a pose was the preservation of the fossil itself. There are some noticeable differences in proportions within pairs of bones on the left and right sides of the skeleton.

The difference in length between the left and right legs influenced the mount pose design to give the skeleton a more natural look.

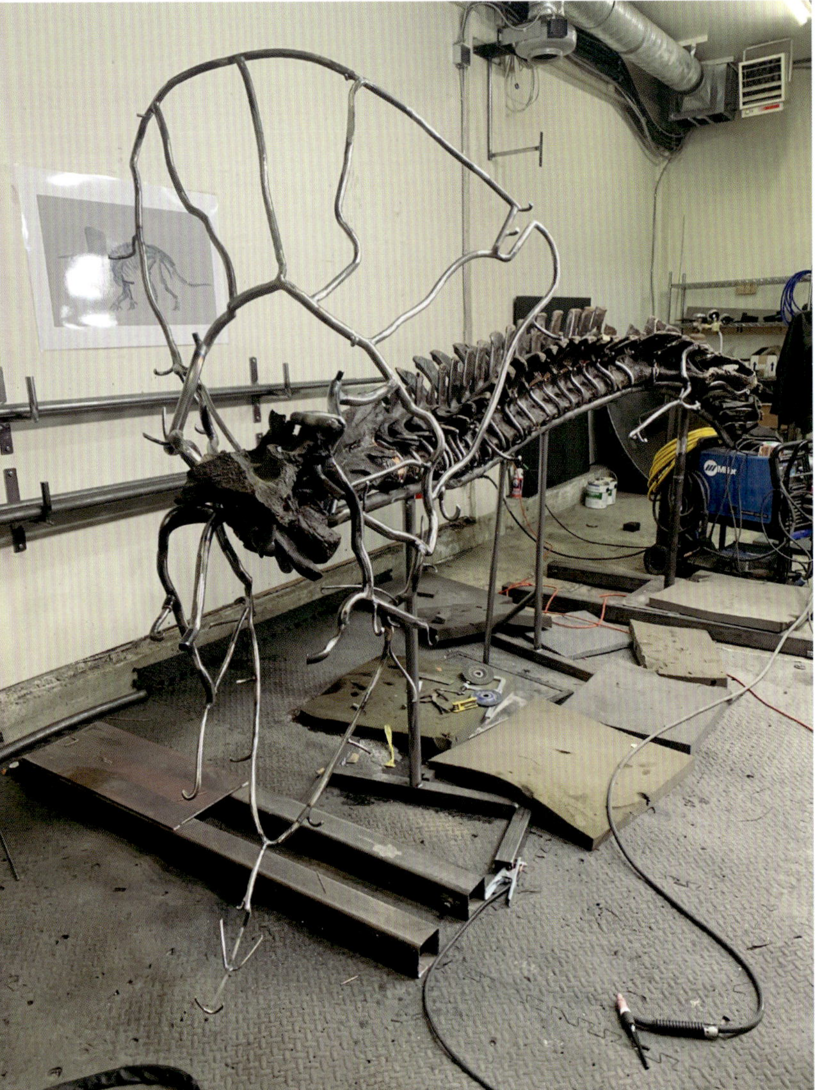

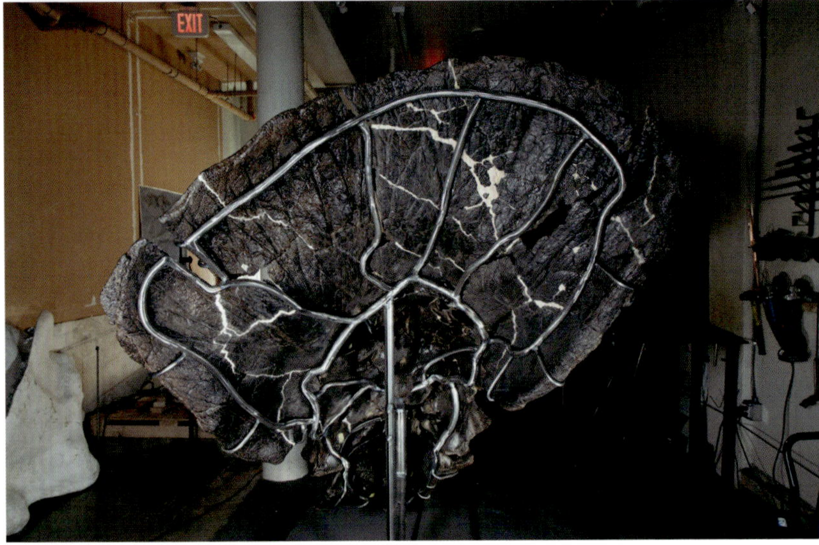
EXIT

Rebuilding *Triceratops*

The right side of the skull is distorted: the frill is squashed on this side, and the brow horn core on the right is flattened and more pointed than the horn core to the left. After it died, the head of the *Triceratops* came to rest on its right side. Over millions of years, the weight of overlying layers of rocks pushing down on the skeleton squashed and slightly stretched the side of the skull it was lying on.

This geologic distortion to the skull bones, particularly those in front of the eye sockets, made it difficult to mount the skull with all its bones exactly lined up and joined together. It was therefore decided to not glue together all the pieces of the skull, which is why there are some big gaps in the front half of the mounted skull, especially on its right side. This was done to try and restore some symmetry to the look of the skull. I also saw an opportunity to CT (computed tomography) scan nearly every part of the skull if the various pieces were kept as small as possible so they could fit through the 'donut ring' of a standard medical CT scanner.

OPPOSITE The complex armature design required artisinal welding skills.

The entire skull was initially mounted on a test armature rig to ensure all the parts fit together properly.

Even more influential on the pose is the 15 percent difference between the length of the left and right thighbones (femora): the left femur is about 103 centimetres long, and the right femur is 87 centimetres long. The right femur, preserved in the rock in a vertical position, was compressed from top to bottom, making it shorter than in life. This left us with a *Triceratops* skeleton in which one leg was a lot shorter than the other! To minimise the impact of this distortion on the 'natural' look of the mounted skeleton, palaeontologist and anatomist Hazel Richards tested how the leg and hip bones would fit together and look if the right leg was posed with its foot raised off the ground. This would make the difference in leg length 'blend in'. Hazel was able to do this in Melbourne thanks to the 3D surface scans of the skeleton that Dino Lab made in Canada and uploaded for access in Australia. Having a virtual 3D model of the skeleton, combined with web meetings of the Museum/Dino Lab team, made it possible to design, review and build a superb skeleton mount in the middle of a pandemic.

The resulting mounted skeleton presents a Triceratops *in mid-stride, walking through the Hell Creek woodland, its head raised and alert as it turns to its left.*

Has it spied some plants to eat over there, perhaps where another *Triceratops* is already plucking leaves? Or has it registered the low-frequency booming calls of an adult *Tyrannosaurus?* Just imagine.

Protecting *Triceratops*

Each *Triceratops* bone was packed in form-fitting expanded foam layers to cushion them and prevent any movement during transit.

Each bone was individually packed in high-density foam layered in custom-built timber shipping crates.

Flying *Triceratops*

On 17th of June 2021, a large truck rolled up to Dino Lab in Victoria, Canada to pick up eight wooden crates: six containing the bones of one *Triceratops* skeleton and two holding a gothic menagerie of steel armature that had been powder-coated in black. The journey to Melbourne had begun.

Once the crates were loaded, the truck drove them to Swartz Bay where they boarded a ferry. The embarked *Triceratops* set sail from Vancouver Island, transited the Strait of Georgia in 95 minutes, and arrived in Vancouver.

While export, import and travel documents awaited approval, a ticket reservation for a flight to Australia had to be confirmed. None of this was simple in a pandemic, where fewer international flights were available. Cargo flights from Canada to Melbourne were cancelled. Weeks passed. Would the *Triceratops* ever get to Melbourne? Should it have been sent by container ship? This transport option was briefly considered, but quickly ruled out—a lot could go wrong on the long sea voyage to Australia.

OPPOSITE
Crates containing the *Triceratops* begin their long journey from Canada to Australia.

Then, at 1:52 pm on the 22nd of July, an Air Canada Boeing 777 airliner roared down the runway and lifted-off from Vancouver International Airport. Onboard was 3163 kilograms of crated, airborne *Triceratops*. Destination: Sydney, Australia. The non-stop flight went smoothly, touching down at 10:08 pm on Friday 23rd of July. The crates were disembarked from the aircraft, transferred to a semitrailer truck on the tarmac and departed Sydney at 12:10am, Saturday 24th of July.

At long last, at 2:00 pm in the afternoon of Monday 27th of July, the Triceratops *arrived at Melbourne Museum.*

The crates were opened, and over the next four months the fossils were unwrapped, sorted, labelled and integrated into the museum's Palaeontology Collection. The fossil bones were now ready to be analysed and photographed. All that remained was for this marvel of nature to be shared with the world.

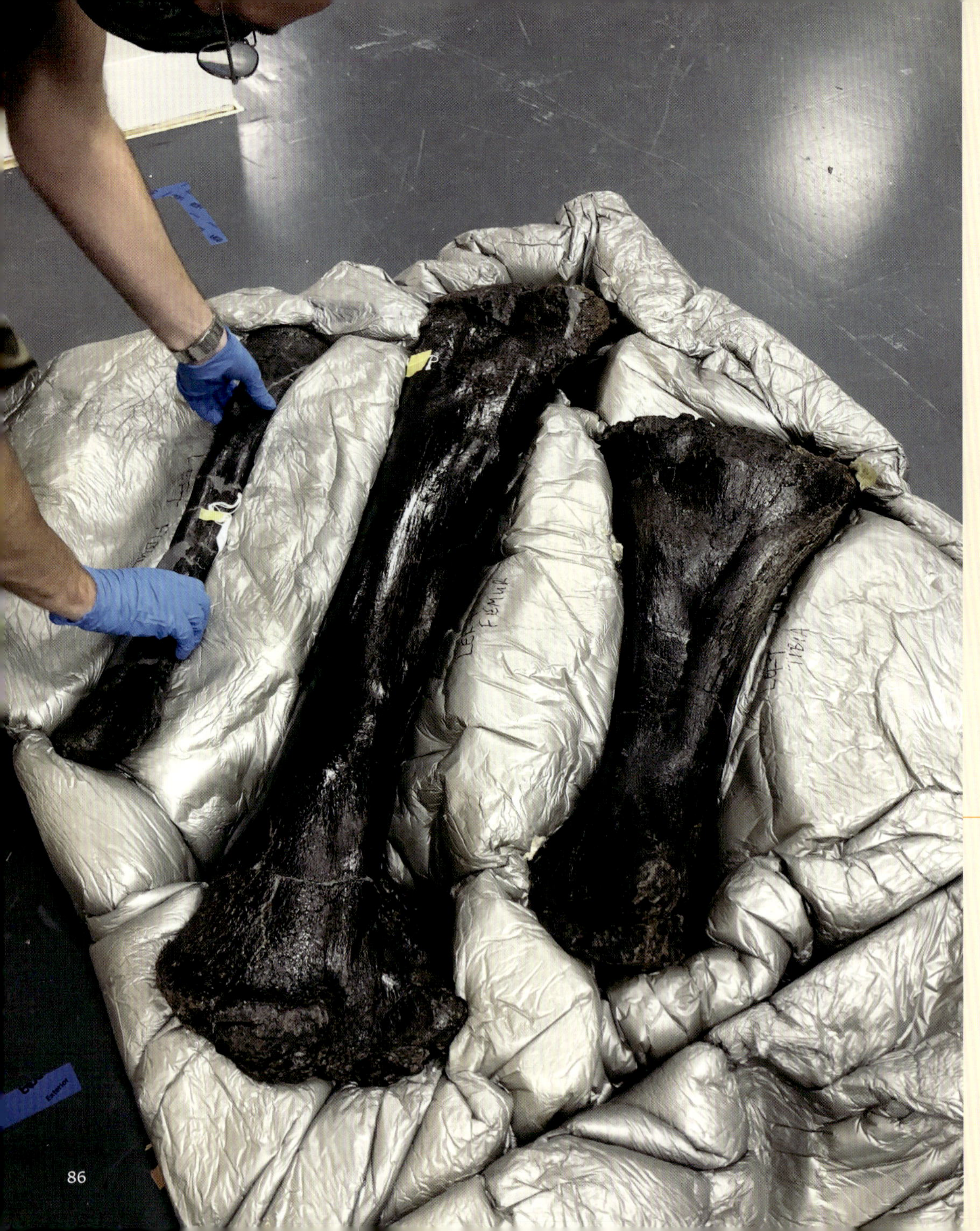

Revealing *Triceratops*

The Melbourne Museum *Triceratops* is one of the most complete and finely preserved *Triceratops* skeletons in the world. How do we know this? At least 267 of the 313 known bones in a *Triceratops* skeleton have been found—so at minimum, 85% of the specimen is complete by bone count. Perhaps more remarkable than this figure is the fact that nearly all the bones identified are intact. The only preserved bones with more than about 25% of their volume missing are the right ischium, right fibula, a right foot (metatarsal) bone and nine of the tail (caudal) vertebrae. Many bone fragments have not yet been identified, and some of these may prove to be parts of bones currently thought to be missing.

After arrival at Melbourne Museum each part of the *Triceratops* skeleton was carefully removed from the shipping crates.

OPPOSITE The skeleton of Horridus is nearly complete, with only those bones shaded dark grey not preserved in the fossil.

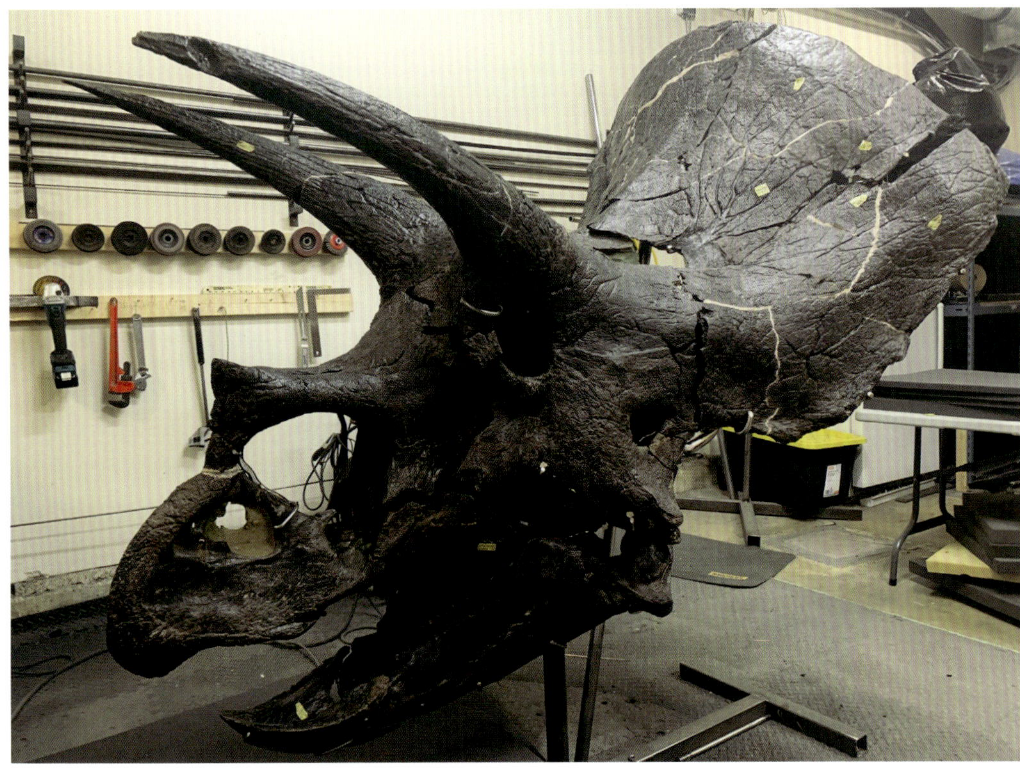

ABOVE The entire skull mounted for the first time.

OPPOSITE Bones or parts of bones not preserved in the skeleton were painted grey to clearly distinguish them from the original fossil that makes up the majority of the *Triceratops*.

The skull on display is complete apart from one or two of the small bones called epiossifications that fit on the rear edge of the frill.

There are not many fossil skulls of Triceratops *in museums that have no areas of major restoration to make them look more complete than they are.*

The length of the skull, measured from the front of the upper beak bone to the rear edge of the frill is about 210 centimetres. The frill is approximately 130 centimetres wide, but this is a bit narrower than it would have been before burial because of the distortion to the fossil. The anatomy of the skull shows that the Melbourne *Triceratops* belongs to the species *Triceratops horridus*, which is the geologically earlier of the two species of *Triceratops*. This is consistent with the geology at the fossil's field site, suggesting it was collected from rocks of the more ancient, lower half of the Hell Creek Formation.

Behind the skull, the Melbourne *Triceratops* skeleton has a complete backbone, intact from the first neck (cervical) vertebra called the atlas all the way back to the smallest vertebra at the tip of the tail—which we now know has 36 vertebrae in it. Although the tail is complete, nine of the vertebrae from the front one-third of the tail are less intact than the others, missing their spines that project up. These nine vertebrae, along with many other smaller bones, had eroded out into the bottom of the pit long before Craig Pfister found the *Triceratops*. Bundles of long ossified tendons (tendons that mineralised into bone) are preserved along the top of the backbone.

Never has a complete backbone of a *Triceratops* been found or placed on public display.

For the first time, the Melbourne *Triceratops* gives an accurate idea of the total length of a *Triceratops* skeleton that is not a composite of more than one individual. Standing in a neutral pose (with the backbone in a straight line), the Melbourne *Triceratops* measures about 6.2 metres long and stands 2.2 metres high (at the top of the backbone). If measured along the curve of the backbone, it is about 6.9 metres long.

The bones from behind the skull (postcranial bones) that were not found are all quite small: chevrons that project down from the tail vertebrae, plus wrist, finger, ankle and foot bones. Why is it that just these few, small bones are missing when nearly all the rest of the skeleton is preserved? More research is needed to answer this question. But evidence from the rocks the *Triceratops* was found in, plus the fossil itself, may provide clues to the circumstances of the *Triceratops's* death and exceptional preservation.

The completely mounted skeleton of *Triceratops horridus*.

How old is Horridus?

It is not yet known exactly how old the Melbourne *Triceratops* was at the time of their death. Other *Triceratops* skulls with similar features—such as total length, brow-horn length and orientation (pointing forward), tightness of sutures between bones and shape of frill edge—have been estimated as adults when they died.

Based on its size alone, the Melbourne *Triceratops horridus* looks like they should be an adult, but it has not been confirmed that they were fully grown at death. Analysis of the internal structure of their bones may provide more evidence on this question, including if the *Triceratops* had stopped growing before they died.

The *Triceratops* was fossilised in fine-grained sandstone, which in the Late Cretaceous was sand deposited on the bottom of a slowly flowing river channel.

The completeness and articulation of the skeleton tells us that the *Triceratops* probably died in the water, or nearby—for example, on a riverbank. If they died on land, they must have been washed into the river soon after death, before decomposition and disarticulation by scavengers (e.g. carnivorous dinosaurs). The carcass may have floated downstream, but sank quickly before breaking apart. We know the *Triceratops* carcass sank to the riverbed with their belly facing down because the fossil was preserved lying on their underside. The carcass settled leaning to their right, with the right arm bent backwards, palm of the hand facing up, and both left and right legs bent at the knees. On the riverbed, the left side of the carcass was breached as the remains were entombed by sand soon after coming to rest. Because the backbone was preserved articulated without collapsing, the body cavity must have been filled with sediment before soft tissues along the back had completely decomposed.

The rapid burial of the skeleton articulated in three dimensions kept the shape of the living animal's body. This was probably an important factor in preserving the Melbourne *Triceratops* skeleton in such rare completeness.

Already, CT scans have revealed that the brain cavity of the Melbourne *Triceratops* may be more intact than in other specimens. The olfactory bulbs of the brain look large—did *Triceratops* have a better sense of smell than we thought? This is just one of many questions about this special fossil (and *Triceratops* in general) that remain to be investigated. How did this particular *Triceratops* die? Are there any signs of disease or injury in their bones? Will we be able to tell if they were a male or a female? How much did they weigh? How did the arms and legs of *Triceratops* work when it was moving, and could it run? What did *Triceratops* eat? How did *Triceratops* evolve to become a dominant dinosaur at the end of the Cretaceous Period?

With their unique preservation and skeletal completeness, Horridus will feature in exciting scientific research exploring these questions and more for decades to come.

Much of what we think we know about this celebrity dinosaur looks set to be changed. Gone for 66 million years, *Triceratops* is back—more real and more alive in our imagination than ever.

Triceratops horridus

Further Reading

Archibald, J. David. 1996. *Dinosaur Extinction and the End of an Era: What the Fossils Say*. Columbia University Press, New York.

Bakker, Robert T. 1986. *The Dinosaur Heresies: A Revolutionary View of Dinosaurs*. Longman Scientific & Technical, Harlow.

Benton, Michael J. 2019. *The Dinosaurs Rediscovered: How a Scientific Revolution is Rewriting History*. Thames & Hudson, London.

Benton, Michael J. 2021. *Dinosaurs: New Visions of a Lost World*. Thames & Hudson, London.

Brett-Surman, M. K., Holtz, Thomas R. and Farlow, James O (editors). 2012. *The Complete Dinosaur*. Second Edition. Indiana University Press, Bloomington.

Brusatte, Stephen L. 2012. *Dinosaur Paleobiology*. Wiley-Blackwell, Chichester.

Brusatte, Stephen L. 2018. *The Rise and Fall of the Dinosaurs: A New History of a Lost World*. Harper Collins, New York.

Dingus, Lowell. 2018. *King of the Dinosaur Hunters: The Life of John Bell Hatcher and the Discoveries that Shaped Paleontology*. Pegasus Books, New York.

Dingus, Lowell and Norell, Mark A. 2021. *The Dinosaur Hunters: The Extraordinary Story of the Discovery of Prehistoric Life*. Welbeck, London.

Dodson, Peter. 1996. *The Horned Dinosaurs: A Natural History*. Princeton University Press, Princeton.

Hatcher, John B., Marsh, Othniel C. and Lull, Richard S. 1907. *The Ceratopsia*. United States Geological Survey Monograph 49:1–300.

Naish, Darren. 2021. *Dinopedia: A Brief Compendium of Dinosaur Lore*. Princeton University Press, Princeton.

Naish, Darren and Barrett, Paul M. 2018. *Dinosaurs: How They Lived and Evolved*. The Natural History Museum, London.

Norell, Mark A. 2019. *The World of the Dinosaurs: The Definitive Illustrated Collection*. André Deutsch, London.

Parker, Tom. 2021. *Saurian: A Field Guide to Hell Creek*. Titan Books, London.

Paul, Gregory S. 2016. *The Princeton Field Guide to Dinosaurs*. Princeton University Press, Princeton.

Ryan, Michael J., Chinnery-Allgeier, Brenda J. and Eberth, David A. (editors). 2010. *New Perspectives on Horned Dinosaurs: The Royal Tyrrell Museum Ceratopsian Symposium*. Indiana University Press, Bloomington.

Sampson, Scott D. 2009. *Dinosaur Odyssey: Fossil Threads in the Web of Life*. University of California Press, Berkeley.

Sax, Boria. 2018. *Dinomania: Why We Love, Fear and Are Utterly Enchanted by Dinosaurs*. Reaktion Books, London.

Glossary

Age of Mammals An alternative name for the Cenozoic Era. This name is based on mammals being the dominant large animals on Earth during this division of the geologic timescale.

amber Fossilised tree resin. Sometimes containing plant and animal remains trapped in the sticky resin when it was soft.

anatomy The shape, size, position and structure of the parts that make up animals and plants.

angiosperms Flowering plants.

armature The three-dimensional framework that supports the bones of a skeleton in a specific pose. Often made of steel.

articulated Connected at joints. Often in reference to a fossil skeleton with the bones connected as they would have been when the animal was alive.

atlas The first neck vertebra of the backbone.

badlands Dry landscape with little vegetation where fossils can be found. In badlands, sedimentary rocks and soils are heavily eroded to form steep gullies and ridges.

biodiversity The variety of organisms in a biological group, or a measure of the richness of life in a particular habitat, ecosystem or area.

bipedal Stance and movement on two (back) legs.

bonebed A layer of concentrated fossil bones in sedimentary rock.

brow horns The pair of horns on the brow, one above each eye socket, of horned dinosaurs.

Campanian The second-last Stage of the Cretaceous Period. Represents the time from about 83.5 to 72 million years ago on the geologic timescale.

caudal vertebra A bone in the tail region of the backbone.

Cenozoic The Earth's current geologic Era. Represents the last 66 million years of the geologic timescale.

ceratopsians Extinct, beaked herbivorous dinosaurs such as the horned dinosaurs that include *Triceratops*.

ceratopsids Large extinct quadrupedal herbivorous dinosaurs with horns and a frill on their skull.

cervical vertebra A bone in the neck region of the backbone.

chevron A small arch-shaped bone with a spine that hangs down from a caudal vertebra in the tail skeleton.

classification The organisation of organisms into groups that are evolutionarily related.

cochlea The part of the inner ear used for hearing.

composite skeleton A skeleton, often a fossil, that is put together using the bones from two or more different individuals.

coprolite Fossilised faeces (dung).

Cretaceous The third and last Period of the Mesozoic Era on the geologic timescale. Lasted from about 145 to 66 million years ago.

crocodylians Crocodiles, alligators, gharials and their fossil relatives.

CT scans Computed tomography scans using X-rays from different angles processed on a computer to make detailed virtual 'slices' and 3D models of the interior of an object, including fossils.

ecosystem The organisms and environment in which the organisms live.

Glossary

endocast An internal cast of a hollow object, such as a braincase.

epijugal A small horn-like bone on the side of the facial region of the skull in ceratopsian dinosaurs.

epiossification A separate bone that projects from the edge of the frill in ceratopsian dinosaurs.

Era A large span of geologic time lasting tens to hundreds of millions of years.

femur The upper leg or thigh bone.

fibula The smaller, outer of the two lower leg or shin bones.

fieldwork In palaeontology, the collection of fossils and other geologic data outdoors.

fossil The preserved remains, impression, or trace of something that was once alive and buried in the geologic past.

frequency The number of regularly repeating waves of a sound that determine its pitch.

genus A unit of biological classification in which species are placed. In the two-part scientific names of organisms, the genus name is the first part of the name for each species. For example, *Triceratops horridus* and *Triceratops prorsus* are two species classified within the genus *Triceratops*.

hadrosaurids Also known as hadrosaurs. Extinct duck-billed herbivorous dinosaurs with flattened beaks. Common in the Late Cretaceous of the northern hemisphere.

hertz The standard unit of frequency.

hindgut fermentation Digestive process in herbivorous animals with a single-chambered stomach.

horn core The bone that is the internal base and support for the horn sheath.

horned dinosaur A member of the extinct horned and frilled herbivorous dinosaur group Ceratopsidae.

ischium One of the pair of bones forming the lower and back part of the pelvis. The plural is ischia.

Jurassic The second Period of the Mesozoic Era on the geologic timescale. Lasted from about 201 to 145 million years ago.

Maastrichtian The last Stage of the Cretaceous Period and Mesozoic Era on the geologic timescale. Lasted from about 72 to 66 million years ago.

Mesozoic The second-to-last Era of Earth's geologic timescale. Lasted from about 252 to 66 million years ago. Includes the Triassic, Jurassic and Cretaceous Periods. Sometimes called the 'Age of Reptiles' or 'Age of Dinosaurs' because the dominant large animals on Earth during this time were dinosaurs and a variety of animals often grouped under the name 'reptiles'.

metatarsal A bone in the foot positioned between the ankle and toe bones.

mudstone A fine-grained type of sedimentary rock that was originally mud or clay.

non-bird dinosaurs All dinosaurs except birds.

olfactory bulb Rounded structures at the front of the brain that receive olfactory (smell) sensations from the nose.

ornithischians Major extinct group of mainly herbivorous dinosaurs. Named for their pelvis structure that looks superficially like that of birds. Includes the well-known *Stegosaurus*, *Ankylosaurus*, hadrosaurs and *Triceratops*. Ornithischians share a unique beak-like bone at the front of the lower jaw called the predentary.

Glossary

ossified tendon A tendon that has transformed into bone.

palaeoecology The study of interactions between organisms, or interactions between organisms and their environments, in the fossil record.

palaeontologist A person who studies fossils. Palaeontology is the study of fossils.

Paleogene The first Period of the Cenozoic Era on the geologic timescale. Lasted from 66 to 23 million years ago.

Period A division of geologic time lasting from just a few million to hundreds of millions of years.

pterosaurs Flying reptiles of the extinct group Pterosauria. Pterosaurs are not dinosaurs, but they are close evolutionary relatives.

quadrupedal Stance and movement on all four limbs (both arms and legs).

rostral The toothless 'beak' bone at the front of the upper jaw, unique to horned dinosaurs.

sacrum The block of fused vertebrae that connects the backbone to the pelvis.

sandstone A type of sedimentary rock made of sand-sized grains of silica minerals like quartz.

sediment Sand, silt or other material, often broken down from sedimentary rocks by weathering and erosion, then transported by water or wind.

sedimentary rock A type of rock formed by the deposition and solidification of sediment.

sinus A sac or cavity in any bone, other tissue or organ in an animal's body.

species The basic unit of biological classification. Defined as a group of organisms that can breed together and produce fertile offspring. This is usually impossible to prove in fossils, so fossil species are often defined as a group of organisms that all share anatomy unique to that group.

Stage A sequence of rock strata that formed in a single age of geologic time lasting anywhere from a few million years to tens of millions of years.

strata Layers of rock (singular: stratum).

suture A joint between two or more bones that does not move.

tectonics Geological processes, like mountain-building, that control the structure of the Earth's crust.

theropods The major group of dinosaurs including all the bipedal predatory dinosaurs of the Mesozoic Era, and birds.

tibia The larger and inner of the two lower leg bones—the shin.

tooth battery Many stacked, interlocking small teeth in rows, which form a long shearing surface in the jaws of horned dinosaurs.

trackway A series of footprints made by an animal.

Triassic The first Period of the Mesozoic Era on the geologic timescale. Lasted from about 252 to 201 million years ago.

vertebra A bone from the backbone. The plural is vertebrae.

Western Interior Seaway A large inland sea that flooded the continent of North America, separating the dry land into western and eastern landmasses during the Cretaceous Period.

Acknowledgements

The *Triceratops* project was made possible with the support of the State Government of Victoria through the Community Support Fund.

Craig Pfister provided essential information on his discovery, fieldwork and preparation of the *Triceratops*. Terry Ciotka and the team from Dino Lab, including Kathryn Abbott, Carly Burbank, Robert Cookson, Blair Holtner, Melissa Kay, Jake Lee, Kalene Lillico, Elaine Vallis, Ry Williams and Nate Williams, are thanked for their sterling preparation, restoration, mounting, documentation and packing of the *Triceratops*. This was a huge collaborative effort sustained over more than 18 months, between opposite hemispheres, in the middle of an unprecedented pandemic. Trevor Dixon Bennett (Kingtide Films) photographed and filmed the *Triceratops* during its time in Canada. Christopher Kukura (Denbigh Fine Art Services), Kingsley Mundey and George Boubeta (IAS Fine Art Logistics) helped enormously with the logistics of packing and transporting the *Triceratops*.

Hazel Richards and James Rule are thanked for their brilliant scientific support and expert contributions to 3D-capturing the *Triceratops*. Hazel Richards provided key anatomical expertise and technical skills in the mounting, digital modelling, and 3D rendering of the *Triceratops* skeleton.

It was a privilege working with palaeoartists Raul Ramos and Rebecca Dart, who produced scientifically credible and beautiful original artwork of the *Triceratops* and its world.

At Museums Victoria, Nancy Ladas, Elizabeth McCartney, Dani Measday, Sarah Babister, Anthony Abell, Wayne Gerdtz, Neville Quick, Oskar Lindenmayer, Rolf Schmidt and Tim Ziegler provided world-class expertise in the management, care, conservation and accessibility of the *Triceratops* as a museum collection item. John Broomfield, Jon Augier, Ben Healley and Rod Start photographed and 3D-captured the specimen.

The American Museum of Natural History, Black Hills Institute of Geological Research, Canadian Museum of Nature, Field Museum of Natural History, Royal Tyrrell Museum of Paleontology, Smithsonian Institution, Yale Peabody Museum of Natural History, Monash University Library and the Paleontological Research Institution are thanked for providing access to images.

Thanks to our Experience Design partners, Grumpy Sailor Creative, who helped bring the world of *Triceratops* to life in the Museum using cutting edge technologies to tell the story in such an immersive way. And their partners Jumbla (3D Animation & Motion Design), Wax Sound Media and RMIT University (Sound Design).

Andrey Atuchin, Julius Csotonyi, Boban Filipovic, James Gurney, Douglas Henderson, Raúl Martín, Mark Witton and Beth Zaiken made their wonderful works of palaeoart available for use in this book.

Palaeontologists who provided valuable information and advice during the preparation of this book include Victoria Arbour, Matthew Carrano, Nathan Carroll, Karen Chin, Philip Currie, Heinrich Mallison, Jordan Mallon, Stephen Poropat and John Scannella. In addition, I thank Matthew Carrano, Philip Currie and Jordan Mallon for reviewing the entire manuscript.

Ursula Smith gave invaluable assistance in sourcing pictures for the book. Gemma Field and Simone Hill are thanked for their design of the book's appearance and layout. I thank Patty Brown and Ella Meave (Museums Victoria Publishing) for their keen interest, management and editorial advice.

Karen, Elsa and Clara were always there as the best, most patient, understanding and enthusiastic support team. Finally, I thank my mother, Christina Fitzgerald, for always fostering my passion, listening to ideas, improving my writing, and reminding me to keep it all in perspective, from the very beginning. This is for you.

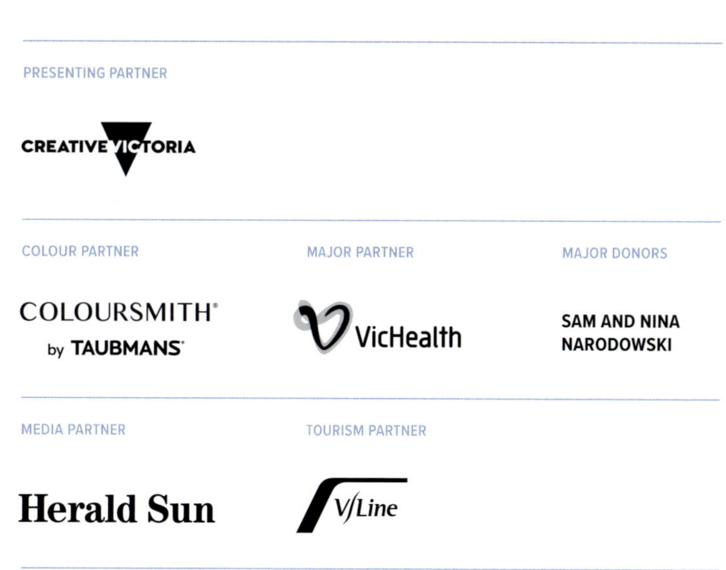

Index

Index

Index

Image Credits

Cover Raul Ramos. Source: Museums Victoria.

Title and chapters 1 to 4 Rebecca Dart. Source: Museums Victoria.

Introducing *Triceratops*

p11 Charles R. Knight. Historic painting of *Triceratops* in its environment, 1901. Courtesy of the Smithsonian Institution. Catalogue number: USNM_V_5799.

p12 Originally published in: Marsh, O.C. 1887. *Notice of new fossil mammals.* American Journal of Science 34:323–324.

p13 Originally published in: Marsh, O.C. 1891. *Restoration of Triceratops*. American Journal of Science 41:339–342.

Originally published in Hatcher, J.B. et al. 1907. *The Ceratopsia*. Monographs of the United States Geological Survey 49: 1-300.

Elisabeth Fulda. *Triceratops*, mounted skeleton, right side view, 1923. AMNH 310443 © American Museum of Natural History.

p14 Ron Blakey. © Ron Blakey. 2014 Colorado Plateau Geosystems Inc. Modified by Erich Fitzgerald.

p15 Heinrich Mallison. Hell Creek Formation, Montana. Source: Museums Victoria.

Evolution of *Triceratops*

p17 *Pachycephalosaurus grangeri* type skull, from Montana. AMNH ID 319456 © American Museum of Natural History.

WehaveaTrex. *Styracosaurus albertensis*. AMNH ID 324091. CC BY 4.0

fig. Andrew A. Farke. *Yinlong downsi*, *Protoceratops andrewsi* and *Stegosaurus stenops*. CC BY 3.0
Raven Amos. *Triceratops prorsus*. CC BY 3.0
Jack Mayer Wood. *Parasaurolophus walkeri*. CC0 1.0
Ian Reid. *Apatosaurus louisae*. CC BY 3.0
Steven Traver. *Gallus gallus*. CC0 1.0

p18 ABelov2014. CC BY 3.0. Source: Deviantart.com

p19 Andrey Atuchin. *Yinlong downsi*, *Psittacosaurus*, *Aquilops*. PaleoNeolitic. *Protoceratops andrewsi.* CC BY 4.0.

p20 John Broomfield. *Protoceratops*. NMV P 207796. Source: Museums Victoria.

Andrey Atuchin. *Protoceratops*.

p21 Andrey Atuchin. *Zuniceratops* resting on riverbank.

p22 Julius Csotonyi. *Anchiceratops ornatus, Arrhinoceratops brachyops, Centrosaurus apertus, Chasmosaurus belli, Diabloceratops eatoni, Eotriceratops xerinsularis, Judiceratops tigris, Kosmoceratops richardsoni, Nasutoceratops titusi, Pachyrhinosaurus lakustai, Pentaceratops sternbergii, Styracosaurus albertensis, Torosaurus latus, Triceratops horridus, Triceratops prorsus,* and *Xenoceratops foremostensis.*

Julius Csotonyi. *Regaliceratops*. Image courtesy of the Royal Tyrrell Museum, Drumheller, AB.

p23 Andrey Atuchin. *Centrosaurus, Chasmosaurus.*

Slate Weasel. Ceratopsid body size chart. Public Domain Mark 1.0. Modified by Erich Fitzgerald.

p24–25 Julius Csotonyi. Niche partitioning among Dinosaur Park Formation dinosaurs. © Canadian Museum of Nature.

p26 Julius Csotonyi. *Regaliceratops*. Image courtesy of the Royal Tyrrell Museum, Drumheller, AB.

p27 Yale Peabody Museum, 1914. *Torosaurus* skull. Courtesy of the Yale Peabody Museum.

p28 Holotype skull in left lateral view of *Triceratops prorsus* [YPM VP 1822]. Courtesy of the Division of Vertebrate Paleontology; Peabody Museum of Natural History, Yale University; peabody.yale.edu. Photography by Vanessa R. Rhue.

Trevor Dixon Bennett. *Triceratops horridus*. Modified by Erich Fitzgerald. Source: Museums Victoria.

p29 Raúl Martín. *Triceratops.*

The World of *Triceratops*

p31 Ron Blakey ©2014 Colorado Plateau Geosystems Inc.

p32–35 Beth Zaiken. Sue's World. © Field Museum, illustration by Blue Rhino Studios.

p36–38 Rebecca Dart. Source: Museums Victoria.

p38–39 Beth Zaiken. Sue's World. © Field Museum, illustration by Blue Rhino Studios.

p40–41 Rebecca Dart. Source: Museums Victoria.

p42 Boban Filipovic.

p43 Rebecca Dart. Source: Museums Victoria.

p45 Douglas Henderson.

Image Credits

The Life of *Triceratops*

p47–48 Raul Ramos. Source: Museums Victoria.

p49 Original Lane skin. Copyright © (2022) Black Hills Institute of Geological Research, Inc.

p50–51 Hazel Richards. Source: MV Sciences.

p52 Raul Ramos. Source: Museums Victoria.

p53 John Broomfield. Source: Museums Victoria.

Originally published in Hatcher, J.B. et al. 1907. *The Ceratopsia*. Monographs of the United States Geological Survey 49: 1-300.

p54 Originally Published in Nabavizadeh, Ali. 2020. *New reconstruction of cranial musculature in ornithischian dinosaurs: implications for feeding mechanisms and buccal anatomy.* The Anatomical Record 303: 347-362.

p55 Raul Ramos. Source: Museums Victoria.

p57 Hazel Richards. Source: MV Sciences.

p58 Douglas Henderson.

p60 Published in Lucas, F. A. 1929. *Animals of the Past: An account of some of the creatures of the ancient world.* American Museum of Natural History Handbook Series no. 4. Seventh edition.

p61 Raul Ramos. Source: Museums Victoria.

p63 James Gurney.

p64 Mark Witton.

p67 Douglas Henderson.

One Special *Triceratops*

p68 Experience Design and Media Production by Grumpy Sailor, image depicts modern day Montana. Source: 3D Animation & Motion Design by Jumbla.

p69 Craig Pfister. Field Journal. Source: Museums Victoria.

p70–71 Heinrich Mallison. Source: Museums Victoria.

p72 Craig Pfister. Horridus skull photo, field map. Source: Museums Victoria.

p73 Hazel Richards. Source: MV Sciences.

p74–75 John Broomfield. Source: Museums Victoria.

p76 Trevor Dixon Bennett. Source: Museums Victoria.

p77 Terry Ciotka. Plaster jacket. Source: Museums Victoria.

Trevor Dixon Bennett. *Triceratops horridus* bones in a field jacket. Source: Museums Victoria

p78 Trevor Dixon Bennett. *Triceratops*: metacarpal and phalanx manual. Source: Museums Victoria

Trevor Dixon Bennett. Kathryn Abbott 3D surface scanning vertebrae of the *Triceratops*. Source: Museums Victoria.

p79 Terry Ciotka. Source: Museums Victoria.

p80 Trevor Dixon Bennett. Ry Williams welding the steel armature for the *Triceratops*, Skull armature, Source: Museums Victoria.

Trevor Dixon Bennett. The partially completed assembly of the Triceratops skull in Dino Labs's workshop in British Columbia, Canada. Source: Museums Victoria

Terry Ciotka. Preparation of *Triceratops* specimen at Dino Lab in Canada. Source: Museums Victoria

p81 Terry Ciotka. *Triceratops horridus,* skull portions. Source: Museums Victoria

p82 Hazel Richards. Source: MV Sciences.

p83 Rod Start. Source: Museums Victoria.

p84 Terry Ciotka. Source: Museums Victoria.

p85 Christopher Kukura. Crates on the airport tarmac in Canada. Source: Museums Victoria.

Terry Ciotka. Crates being loaded onto a truck in Canada. Source: Museums Victoria.

p86 Erich Fitzgerald. Source: Museums Victoria.

p87 Hazel Richards. Source: MV Sciences.

p88 Terry Ciotka. Source: Museums Victoria.

p89 Trevor Dixon Bennett. Source: Museums Victoria.

p90 Rod Start. Source: Museums Victoria.

p93 Raul Ramos. Source: Museums Victoria.